International Union of Crystallography
Commission on Crystallographic Computing

WORLD LIST
OF
CRYSTALLOGRAPHIC COMPUTER PROGRAMS

SECOND EDITION

edited by

DAVID P. SHOEMAKER

(Massachusetts Institute of Technology, Cambridge, Mass., U.S.A.)

1966

Copies of this World List have been distributed free of charge to all subscribers to *Acta Crystallographica*, a journal of the International Union of Crystallography. Additional copies can be obtained from A. Oosthoek's Uitgevers Mij. N.V., Domstraat 11-13, Utrecht, The Netherlands, at the price of 10 Netherlands Guilders (U.S. $ 3.00 or U.K. £ 1 at the present rates of exchange) per copy, including postage for shipment by surface mail. In the event of foreign exchange difficulties, UNESCO coupons will be accepted. Orders can also be placed with Polycrystal Book Service, P.O. Box 11567, Pittsburgh, Pa. 15238, U.S.A.; or with any bookseller.

ISBN 978-90-277-9034-7   ISBN 978-1-4684-8477-9 (eBook)
DOI 10.1007/978-1-4684-8477-9

Bronder - Offset Rotterdam
The Netherlands

I U CR WORLD LIST OF CRYSTALLOGRAPHIC COMPUTER PROGRAMS

SECOND EDITION
JULY 1, 1966

INTERNATIONAL UNION OF CRYSTALLOGRAPHY
COMMISSION ON CRYSTALLOGRAPHIC COMPUTING

GENERAL DESCRIPTION

1. THE FORMAT OF THIS LIST IS THAT OF THE FIRST EDITION, AND OF PREVIOUS LISTS OF THE AMERICAN CRYSTALLOGRAPHIC ASSOCIATION. IT IS BASED ON THE USE OF STANDARD 80-COLUMN IBM CARDS.

2. NORMALLY, EACH PROGRAM LISTED SHOULD BE A DISCRETE ENTITY CAPABLE OF BEING RUN BY ITSELF (ALTHOUGH IT MAY OR MAY NOT NORMALLY BE USED IN CONJUNCTION WITH ONE OR MORE OTHER PROGRAMS). HOWEVER, A PACKAGE OF MINOR PROGRAMS OR ROUTINES EACH OF WHICH HAS LITTLE IMPORTANCE OUTSIDE THE PACKAGE SHOULD PREFERABLY BE LISTED AS A SINGLE ENTRY.

3. A PROGRAM SHOULD NOT BE LISTED UNLESS IT HAS RUN SUCCESSFULLY, ALTHOUGH IT NEED NOT BE DEBUGGED IN ALL RESPECTS. PREFERENCE FOR LISTING IS FOR PROGRAMS WHICH HAVE SOME PROMISE OF BEING USEFUL TO OTHERS AND FOR WHICH SUITABLE WRITE-UPS ARE, OR WILL PROBABLY BECOME, AVAILABLE.

4. EACH PROGRAM IS TO BE REPRESENTED BY AN ENTRY CONSISTING OF --

    A. A TITLE COMPRISING A SINGLE IBM CARD (80 COLUMNS) CONTAINING THE ESSENTIAL INFORMATION PERTAINING TO THE PROGRAM IN A VERY COMPRESSED FORM, AND

    B. AN (OPTIONAL) ABSTRACT OF NOT MORE THAN ABOUT 50 WORDS, WRITTEN BY THE AUTHOR(S) OF THE PROGRAM, ON ADDITIONAL IBM CARDS. THIS SHOULD GIVE WHATEVER CAN BE BRIEFLY STATED REGARDING TYPE OF APPLICATION, GENERALITY, VERSATILITY, SPEED, STORAGE REQUIREMENTS, AND ACCURACY. REFERENCES TO OTHER PROGRAMS THAT CAN BE USED IN CONJUNCTION WITH IT ARE VALUABLE.

5. THE I U CR WORLD LIST OF COMPUTER PROGRAMS WILL CONSIST PRIMARILY OF THE TITLE AND ABSTRACT CARDS ARRANGED ACCORDING TO MACHINE TYPE AND FUNCTION.

6. TO ASSIST IN THE USE OF THE PROGRAM LIST, A LISTING OF ABBREVIATIONS AND AN AUTHOR INDEX WILL BE MADE AVAILABLE FROM PUNCHED CARD LISTINGS. THE EXTREME DEGREE OF COMPRESSION REQUIRED ON THE TITLE CARD MAY REQUIRE THAT EVEN AUTHORS NAMES BE ABBREVIATED. IN SUCH CASES THE AUTHOR INDEX WILL CROSS-REFERENCE THE ABBREVIATED NAME TO THE CORRECTLY SPELLED NAME. ASIDE FROM SUCH CROSS REFERENCES, EACH ENTRY IN THE AUTHOR INDEX WILL CONSIST OF SURNAME, INITIALS, AND MAILING ADDRESS, ON A SINGLE IBM CARD.

7. THE TITLE, ABSTRACT, AND AUTHOR-INDEX CARDS SHOULD BE PREPARED BY THE PROGRAMMERS THEMSELVES WHENEVER POSSIBLE, IN ACCORDANCE WITH THE FORMAT GIVEN BELOW. IF THIS IS NOT POSSIBLE, THE INFORMATION SHOULD BE SUBMITTED ON PROGRAM LIST FORMS, AVAILABLE ON APPLICATION.

8. THE INFORMATION SHOULD BE KEPT AS CURRENT AND COMPLETE AS POSSIBLE BY FRESH INFORMATION FROM PROGRAMMERS, FOR FUTURE EDITIONS. ESPECIALLY, THE STATUS AND AVAILABILITY INFORMATION IN COLS. 78 - 80 SHOULD BE COMPLETE AND UP-TO-DATE. OBSOLETE PROGRAMS, AND PROGRAMS THAT HAVE NO APPRECIABLE PROSPECT OF BECOMING AVAILABLE OR OF BEING USEFUL TO OTHERS, SHOULD BE WITHDRAWN.

9. A SUBSTANTIAL REVISION OR MODIFICATION OF A PROGRAM SHOULD BE TREATED

AS A NEW PROGRAM WITH A NEW ENTRY IN THE PROGRAM LIST AND A NEW ACCESSION NUMBER. THE OLDER PROGRAM SHOULD BE WITHDRAWN IF MADE OBSOLETE BY THE REVISION.

10. THE LANGUAGE OF THE TITLE CARD SHOULD BE ENGLISH. FOR ABSTRACT CARDS ENGLISH IS PREFERABLE THOUGH NOT NECESSARY. HOWEVER, ONLY THE ROMAN ALPHABET WITH LETTERS AND CHARACTERS AVAILABLE ON ENGLISH-LANGUAGE IBM MACHINES MAY BE USED.

1234567890 ABCDEFGHIJKLMNOPQRSTUVWXYZ .,=+-/()*

11. PROGRAMS SHOULD BE LISTED UNDER PROGRAMMING LANGUAGES (PSEUDO-MACHINES) SUCH AS ALGOL, FORTRAN, ETC., RATHER THAN UNDER THE ACTUAL MACHINES, WHEN THEY ARE MACHINE-INDEPENDENT OR LARGELY SO, BUT THE MACHINE ON WHICH COMPILED AND TESTED SHOULD BE CITED.

12. CERTAIN CRYSTALLOGRAPHIC PROGRAM SYSTEMS, AND INDIVIDUAL PROGRAMS BELONGING TO THEM, ARE LISTED IN A SEPARATE CATEGORY (THE FIRST CATEGORY IN THE LIST).

13. THIS EDITION CONTAINS LISTINGS FOR 697 PROGRAMS, OF WHICH 311 ARE FROM THE FIRST EDITION AND 386 ARE NEW. A FEW OF THE LISTINGS IN THE FIRST EDITION HAVE BEEN REVISED FOR THE NEW EDITION, AND A FEW HAVE BEEN TRANSFERRED FROM ONE MACHINE TO ANOTHER OR TO FORTRAN. LISTINGS IN THE FIRST EDITION BUT NOT IN THE SECOND ARE
    A. THOSE SUGGESTED FOR WITHDRAWAL BY THEIR AUTHORS,
    B. THOSE (WITH FEW EXCEPTIONS) WITH NO STATUS AND AVAILABILITY INFORMATION IN COLS. 78 - 80 OF THE TITLE CARD, OR STATED IN COL. 80 AS NOT AVAILABLE, AND
    C. THOSE FOR CERTAIN OLD MACHINES JUDGED TO BE OBSOLETE, MOST PROGRAMS FOR WHICH HAVE BEEN WITHDRAWN BY THEIR AUTHORS AND NO NEW PROGRAMS FOR WHICH HAVE BEEN SUBMITTED.

                                                                                DAVID P. SHOEMAKER, EDITOR
                                                                                   MASS. INST. OF TECH.
                                                                            CAMBRIDGE, MASS. 02139, U.S.A.

I U CR COMMISSION ON CRYSTALLOGRAPHIC COMPUTING
    D. W. J. CRUICKSHANK, CHAIRMAN
    W. R. BUSING
    TH. HAHN
    A. LINEK
    M. A. PORAY-KOSHITS
    D. P. SHOEMAKER
    Y. TAKEUCHI

THE REGIONAL AND MACHINE CORRESPONDENTS WHO AIDED IN THE COLLECTION OF THE PROGRAM LISTINGS ARE NAMED BELOW. ACCOMPANYING THEIR NAMES ARE THEIR ASSIGNED RANGES OF ACCESSION NUMBERS.

    1 - 2999   W. R. BUSING (CANADA, U. S. A.)
                        OAK RIDGE NAT. LAB., P. O. BOX X, OAK RIDGE, TENNESSEE, USA
                FOR FIRST EDITION --
                D. SAYRE
                      IBM WATSON RESEARCH CTR., P.O. BOX 218, YORKTOWN HEIGHTS,
                      NEW YORK 10598, USA
    3001 - 3499 D. ROGERS (ZEBRA, ATLAS, BRITISH COMMONWEALTH)
                      IMPERIAL COLLEGE, UNIV. OF LONDON, LONDON S.W. 7, ENGLAND
    3501 - 3999 D. W. J. CRUICKSHANK (KDF9)
                      CHEM. DEPT., THE UNIVERSITY, GLASGOW W. 2, SCOTLAND
    4001 - 4499 J. ROLLETT (DEUCE, MERCURY)
                      COMPUTING LAB., U. OF OXFORD, SOUTH PARKS RD., OXFORD,ENGLAND
    5001 - 5999 T. HAHN (AUSTRIA, GERMANY, SWITZERLAND)
                      INST. FUR KRISTALLOGRAPHIE, HOCHSCHULE AACHEN,
                      51 AACHEN, TEMPLERGRABEN 55, WEST GERMANY
    6001 - 6499 S. ASBRINK (SCANDINAVIA)
                      INST. OF INORG. AND PHYS. CHEM., U. OF STOCKHOLM,

	KUNGSTENGATAN 45, STOCKHOLM, SWEDEN
6501 - 6999	A. LINEK (BULGARIA, CZECHOSLOVAKIA, GERMAN DEMOCRATIC REPUBLIC, HUNGARY, POLAND, ROUMANIA)
	INST. OF TECHN. PHYS., ACAD. OF SCI., CUKROVARNICKA 10, PRAGUE, CZECHOSLOVAKIA
7001 - 7499	M. TOURNARIE (FRANCE, PORTUGAL, SPAIN)
	40, RUE BANGEON, MASSY, 78, FRANCE
	FOR FIRST EDITION --
	P. MERIEL, C. E. N. SACLAY, B. P. 2, GIF-SUR-YVETTE, SEINE-ET-OISE, FRANCE
7501 - 7999	C. PANATTONI (ITALY, YUGOSLAVIA)
	IST. DI CHIMICA ORGANICA, U. OF PADOVA, PADOVA, ITALY
8001 - 8499	MRS. E. W. M. RUTTEN-KEULEMANS (BELGIUM, NETHERLANDS)
	CENTRAAL REKENINST., UNIV. OF LEIDEN, LEIDEN, THE NETHERLANDS
	FOR FIRST EDITION --
	D. W. SMITS, REKENCENTRUM DER UNIVERSITEIT, APPELSTRAAT 11, GRONINGEN, NETHERLANDS
8501 - 8999	Y. TAKEUCHI (JAPAN)
	DEPT. OF MINERALOGY, U. OF TOKYO, MOTOFUJIMACHI, BUNKYOKU, TOKYO, JAPAN

THE PUNCHED CARD WORK FOR THIS EDITION WAS DONE MAINLY AT THE M. I. T. COMPUTATION CENTER, MASS. INST. OF TECH., CAMBRIDGE, MASS. 02139, U. S. A.

FORMAT

TITLE CARD

COLS. 1 - 4   PROGRAM ACCESSION NUMBER, TO BE ASSIGNED BY THE EDITOR. PROGRAMS ARE NUMBERED SERIALLY IN CHRONOLOGICAL ORDER OF RECEIPT BY THE EDITOR. TO AVOID CONFUSION, ACCESSION NUMBERS OF OBSOLETED PROGRAMS WILL NOT BE RE-USED. THE UNITS PLACE OF THE NUMBER SHOULD BE IN COL. 4.

COLS. 6 - 13   MACHINE TYPE, BY CODE NAME OR NUMBER. MAY BE FOLLOWED BY ABBREVIATIONS INDICATING SIZE OF STORAGE, MODIFICATIONS, ETC., AND/OR LANGUAGE IN WHICH PROGRAM WAS WRITTEN.

ALTERNATIVELY, FOR A PROGRAM WRITTEN IN A MACHINE-INDEPENDENT MANNER IN A LANGUAGE SUCH AS FORTRAN OR ALGOL, THE LANGUAGE NAME MAY TAKE THE PLACE OF THE MACHINE NAME. IN SUCH A CASE IT IS ASSUMED THAT THE USER MUST ORDINARILY COMPILE THE PROGRAM FOR HIS OWN MACHINE. THE ABSTRACT SHOULD STATE THE TYPE OF MACHINE ON WHICH TESTED, AND INDICATE THE AMOUNT OF STORAGE REQUIRED.

COLS. 15 - 22   FUNCTION OF PROGRAM, BY CODE ABBREVIATIONS. SEE LIST OF ABBREVIATIONS.

COLS. 24 - 38   AUTHORS, PROGRAMMERS, ETC. BECAUSE OF SPACE LIMITATIONS MORE THAN ONE OR TWO NAMES SHOULD BE AVOIDED IF THIS CAN BE DONE WITHOUT INJUSTICE. ONLY SURNAMES SHOULD BE GIVEN EXCEPT WHEN USE OF AN INITIAL IS NECESSARY TO AVOID CONFUSION.

IT MAY BE NECESSARY TO ABBREVIATE SURNAMES. IN THIS CASE CROSS REFERENCES SHOULD BE ENTERED INTO THE AUTHOR INDEX.

TWO OR MORE SURNAMES SHOULD BE SEPARATED BY PUNCTUATION (COMMA, SLASH, OR ASTERISK) BUT NO SPACES. WHERE INITIALS ARE NEEDED THEY SHOULD FOLLOW THE SURNAME, SEPARATED BY SPACES BUT NO PUNCTUATION.

WHERE A SLASH (/) APPEARS BETWEEN TWO NAMES, THE NAMES AFTER THE SLASH ARE THE NAMES NOT OF AUTHORS OR PROGRAMMERS BUT

THOSE OF PERSONS PRESENTLY IN CONTROL OF THE PROGRAM, TO WHOM INQUIRIES REGARDING THE PROGRAM SHOULD BE ADDRESSED.

WHERE AN ASTERISK (*) APPEARS AFTER THE NAME OF AN AUTHOR OR PROGRAMMER, IT INDICATES THAT THAT PERSON, RATHER THAN THE FIRST-NAMED AUTHOR OR PROGRAMMER, IS THE PERSON TO WHOM INQUIRIES SHOULD BE SENT.

THE NAMES OF AUTHORS MAY BE PERMITTED TO EXTEND BEYOND COL. 38 AND ENCROACH ON THE NEXT FIELD IF ABSOLUTELY NECESSARY. IN THIS CASE THE LAST NAME GIVEN SHOULD BE FOLLOWED BY A SLASH TO SEPARATE IT FROM WHAT FOLLOWS.

COLS. 40(41) - 76   IDENTIFICATION AND COMMENTS. IF AN IDENTIFICATION CODE NAME OR NUMBER EXISTS, IT SHOULD BE GIVEN STARTING IN COLUMN 40 AND FOLLOWED BY A SLASH (/). FURTHER COMMENTS BEGIN IMMEDIATELY AFTER THE SLASH, WITHOUT A SPACE BEFORE.

IF THERE IS NO IDENTIFICATION CODE NAME OR NUMBER, COLUMN 40 SHOULD BE LEFT BLANK AND THE COMMENTS SHOULD BEGIN IN COLUMN 41.

COMMENTS SHOULD BE ABBREVIATED AS MUCH AS POSSIBLE AND MUST NOT EXTEND BEYOND COLUMN 76.

COL. 78   STATUS OF PROGRAM OPERABILITY AND AVAILABILITY OF PROGRAM CODE (I.E., ACTUAL LISTING OF PROGRAM IN AUTOMATIC AND/OR SYMBOLIC AND/OR MACHINE LANGUAGE, ADEQUATE FOR TROUBLE-SHOOTING AND MAKING MODIFICATIONS)

    L   WELL CHECKED OUT, PROGRAM CODE AVAILABLE
    M   WELL CHECKED OUT, PROGRAM CODE NOT AVAILABLE
    N   OPERABLE BUT NOT WELL CHECKED OUT

COL. 79   STATUS OF PROGRAM WRITE-UP

    W   COMPLETE WRITE-UP IN FINAL FORM AVAILABLE
    P   ABBREVIATED OR PROVISIONAL WRITE-UP AVAILABLE
    N   NO WRITE-UP AVAILABLE

COL. 80   STATUS OF AVAILABILITY OF PROGRAM IN WORKING FORM (PUNCHED CARDS OR TAPE, PLUS WRITE-UP AND/OR PROGRAM CODE IF SO INDICATED IN 78 AND/OR 79)

    S   AVAILABLE THROUGH MACHINE-USER ORGANIZATION (E. G., SHARE)
    M   AVAILABLE THROUGH MANUFACTURER OF MACHINE
    A   AVAILABLE DIRECTLY FROM AUTHOR OR AUTHORS INSTITUTION
    X   PROGRAM IS OF A SPECIAL OR LOCAL NATURE, CONDITIONALLY AVAILABLE
    N   NOT AVAILABLE AT PRESENT, PROBABLY AVAILABLE AT LATER DATE

ABSTRACT CARDS

COLS. 1 - 4   PROGRAM ACCESSION NUMBER, THE SAME AS ON THE TITLE CARD.

COL. 5   SUCESSIVE CARDS OF AN ABSTRACT ARE LETTERED SERIALLY A, B, C, D, . . .

COLS. 8 - 80   (73 CHARACTERS AND SPACES PER CARD) FOR THE CONTENT OF THE ABSTRACT, WHICH SHOULD BE LIMITED TO NO MORE THAN ABOUT 50 WORDS.

AUTHOR INDEX CARDS -- BEGIN IN COL. 1, AND GIVE
SURNAME, INITIALS, AND MAILING ADDRESS, E.G. --

COULTER, C. L., M. R. C. UNIT, CAVENDISH LABORATORY, CAMBRIDGE, ENGLAND

ALL NAME ABBREVIATIONS SHOULD BE EXPLAINED ON ADDITIONAL CARDS, E. G.,

CLTR, SEE COULTER, C. L.

ABBREVIATIONS

FUNCTION ABBREVIATIONS

A	ACENTRIC (NON-CENTROSYMMETRIC) ONLY
ABS	ABSORPTION CORRECTIONS
A-C TEST	ACENTRIC/CENTRIC STAT. TEST
AT	ANISOTROPIC TEMPERATURE FACTORS CAN BE TREATED
BL	BEEVERS-LIPSON TYPE OF (FR) CALCULATION
BLK	SMALL BLOCK OF POINTS, AS IN NEIGHBORHOOD OF FOURIER PEAK
C	CENTRIC (CENTROSYMMETRIC) ONLY
CONTOUR	PRODUCES CONTOURED FOURIER MAPS OR SECTIONS
D	INTERPLANAR SPACINGS
DF	DIFFERENTIAL FOURIER
DEBYE	DEBYE TEMPERATURE
DIR	DIRECT METHODS OF SIGN OR PHASE DETERMINATION
DISPERSN	ANOMALOUS DISPERSION (CORRECTIONS, ETC.)
DM	DIAGONAL (OR MAINLY DIAGONAL) MATRIX
DP	DATA PROCESSING
E	ERROR, STANDARD DEVIATIONS, ETC.
EL	ELECTRON (DIFFRACTION)
EQ	EQUI-INCLINATION GEOMETRY
EXT	EXTINCTION (OR EXT. CORRECTION)
FM	FULL MATRIX
FPK	FOURIER PEAK SHAPE ANALYSIS
FR	FOURIER SYNTHESIS (2 OR 3 DIM)
FT	FOURIER TRANSFORM
GEOM	MOLECULAR GEOMETRY
H	INDEX (HKL) GENERATION
ICA	INDIVIDUAL ATOMIC CONTRIBUTIONS (TO SF) AVAILABLE
ID	INTERATOMIC DISTANCES (AND USUALLY BOND ANGLES)
IT	INPUT
LC	LATTICE CONSTANTS
LF	LITTLE F - FORM FACTORS
LP	LORENTZ-POLARIZATION CORRECTION
LS	LEAST SQUARES REFINEMENT (3D UNLESS OTHERWISE STATED). SEE ALSO SFLS.
N	NEUTRON (DIFFRACTION)
OT	OUTPUT
P	POWDER CALCULATIONS
PARAM	PARAMETER(S)
PATSUP	PATTERSON SUPERPOSITION, MINIMUM FUNCTION, ETC.
PI	POWDER INDEXING
PK	PEAK
PLANE	LEAST-SQUARES PLANE THROUGH GROUP OF ATOMS
PREC	PRECESSION
PRJCTN	PROJECTION
PROF	PROFILE
R	R FACTOR, AGREEMENT FACTOR
ROTF	ROTATION FUNCTION
ROTN(FN)	ROTATION FUNCTION
RB	RIGID BODY
S	SPECIAL
SCL	SCALE FACTOR
SD	STANDARD DEVIATIONS
SF	STRUCTURE FACTORS
SFLS	STRUCTURE FACTORS AND LEAST SQUARES REFINEMENT
SIGN	MANIPULATION OF STRUCTURE FACTOR SIGNS (NOT SAME AS DIR)
SPC	SPECTROMETER SETTINGS
SPEC	SPECIAL
STAT	STATISTICS, PERTAINING PARTICULARLY TO SF
STEREO	CALC OF STEREO STRUCTURE PICTURES
SYSTEM	CRYSTALLOGRAPHIC PROGRAM SYSTEM

```
TF         TEMPERATURE FACTOR
THETA      BRAGG ANGLE
WEIS       WEISSENBERG
2          TWO-DIMENSIONAL
3          THREE-DIMENSIONAL
2C,3C,4C   2-, 3-, 4-CIRCLE (DIFFRACTOMETER)
```

MACHINE ABBREVIATIONS

```
ATLAS      FERRANTI ATLAS
BULL GET   BULL TYPE GAMMA ET (FRANCE)
BULLGAET   BULL TYPE GAMMA AET (FRANCE)
B5000      (NO DEFINITION RECEIVED)
CAB 500    SEA TYPE CAB 500 (FRANCE)
CEP        CALCOLATRICE ELETTRONICA PISANA  (PISA UNIVERSITY, PISA, ITALY)
DASK       (DENMARK)
ER 56      STANDARD-ELECTRIC-LORENZ ER 56 (GERMANY)
FACIT      FACIT EDB  (SWEDEN)
KDF9       ENGLISH ELECTRIC KDF9
LGP30      ROYAL-MCBEE
MERCURY    FERRANTI MERCURY (BRITAIN)
M1B        MUSASHINO 1B, SOMEWHAT SIMILAR TO PC1 (TOKYO, JAPAN)
NE803      NATIONAL ELLIOTT NE 803
PC1        PARAMETRON COMPUTER, DEPT. OF PHYSICS, UNIV. OF TOKYO, TOKYO, JAPAN
PDP6       DIGITAL EQUIPMENT CORP. PDP6
SILLIAC    SILLIAC (AUSTRALIA)
URAL 1     (USSR)
WGMATIC    WEGEMATIC 1000  (SWEDEN)
X 1        N. V. ELECTROLOGICA, THE HAGUE, NETHERLANDS
ZEBRA      STANTEC ZEBRA, STANDARD TELEPHONES AND CABLES LTD, ENFIELD, ENGLAND
Z22R       ZUSE 22R (GERMANY)
Z23A       ZUSE 23A
1103       REMINGTON RAND UNIVAC 1103
1107       REMINGTON RAND UNIVAC 1107
1410       IBM 1410
1604       CONTROL DATA 1604
1620       IBM 1620
2002       SIEMENS 2002 (GERMANY)
205        BURROUGHS 205
220        BURROUGHS 220
360        IBM SYSTEM 360
5090H      OKIDAC-5090H (JAPAN)
6001       OLIVETTI ELEA 6001.  K INDICATES 10000 DIGITS HIGH SPEED STORAGE.
650        IBM 650
704        IBM 704
7040       IBM 7040
7044       IBM 7044
7070       IBM 7070
7074       IBM 7074
709        IBM 709
7090       IBM 7090
7094       IBM 7094
7094DC     IBM 7094 DIRECT COUPLED
NOTE --  IBM 709, 7090, 7094 (IN ORDER OF INCREASING SPEED) ARE LARGELY
         INTERCHANGEABLE, AND MACHINE-LANGUAGE PROGRAMS ARE READILY
         INTERCONVERTABLE.  THESE MACHINES ARE GROUPED TOGETHER IN THE
         PROGRAM LISTINGS.
```

PROGRAMMING LANGUAGES (PSEUDO-MACHINES), USUALLY COMPILABLE ON SEVERAL MACHINE
    TYPES), WHICH MAY BE CITED IN LIEU OF MACHINE IF SOURCE PROGRAM IS LARGELY
    MACHINE INDEPENDENT (HOWEVER, ABSTRACT CARDS SHOULD CONTAIN REFERENCE
    TO MACHINE ON WHICH COMPILED, PREFERABLY WITH INDICATION OF STORAGE
    REQUIREMENTS) --

```
ALGOL      ALGORITHMIC LANGUAGE
```

```
ALGOL60   DITTO (1960 VERSION)
ATLAS F   WRITTEN IN A FORM OF FORTRAN (HARTRAN) FOR THE ATLAS COMPUTER
               (HOWEVER, PROGRAMS IN ATLAS F ARE LISTED UNDER ATLAS MACHINE)
AUTOCODE
EMA       EXTENDED MERCURY AUTOCODE
EXCHLF    (SEE ATSYS)
FORTRAN   FORMULA TRANSLATION (IBM)
FORTRAN2  FORTRAN II
FORTRAN4  FORTRAN IV
HARTRAN   SEE  ATLAS F
MAD       MICHIGAN ALGORITHMIC DECODER (RELATED TO ALGOL)
NP/I      PROGRAMMING LANGUAGE TO BE IMPLEMENTED UNDER IBM SYSTEM 360

MODIFIERS

  AGL     WRITTEN IN ALGOL (60), COMPILED ON MACHINE INDICATED BY PRIOR ABBREV.
  AUC     WRITTEN IN AUTOCODER (7070), COMPILED ON MACHINE INDIC. BY PRIOR ABBREV
  AUT     WRITTEN IN AUTOCODE, COMPILED ON MACHINE INDICATED BY PRIOR ABBREV.
  FTN     WRITTEN IN FORTRAN, COMPILED ON MACHINE INDICATED BY PRIOR ABBREV.
  FT2     WRITTEN IN FORTRAN II, COMPILED ON MACHINE INDICATED BY PRIOR ABBREV.
  8K      INDICATED MACHINE MUST HAVE AT LEAST 8192 WORDS HIGH-SPEED STORAGE
  4K, 32K    MUST HAVE AT LEAST 4096, OR 32,768 WORDS, ETC.

CRYSTALLOGRAPHIC PROGRAM SYSTEMS

ATSYS     ATLAS CRYSTALLOGRAPHIC SYSTEM, COMPILED AT IMPERIAL COLLEGE, LONDON.
             DEVELOPED MAINLY BY M. G. B. DREW.
X-RAY 63  SYSTEM OF CRYSTALLOGRAPHIC PROGRAMS DEVELOPED BY J. STEWART, NOW AT
             UNIV. OF MARYLAND.  CONTAINS INDIVIDUAL PROGRAMS WRITTEN BY MANY
             OTHERS.  INDIVIDUAL PROGRAMS ARE LISTED UNDER 7090/7094 AS
             WELL AS UNDER THE X-RAY 63 SYSTEM.



                              AUTHOR/PROGRAMMER INDEX

ABRAHAMS, S. C., BELL TELEPHONE LABORATORIES, MURRAY HILL, NJ
ABRAHAMSSON, S., INST. OF MEDICAL BIOCHEMISTRY, UNIV. OF GOTEBORG, SWEDEN
ABRAHMS, SEE ABRAHAMS, S. C.
ABRHMSN, SEE ABRAHAMSSON, S.
AHMED, F.R., NATIONAL RESEARCH COUNCIL, OTTAWA 2, ONTARIO, CANADA
ALBANO, V., IST. CHIM. GENERALE, VIA C. SALDINI 50, MILANO, ITALY
ALBA, SEE ALBANO, V.
ALDEN,R.    UNIVERSITY OF CALIFORNIA AT SAN DIEGO LA JOLLA CALIF. USA
ALLEN, S., METALLURGY DEPT., MIT, CAMBRIDGE, MASS.
AMMON,H.    UNIVERSITY OF WASHINGTON   SEATTLE, WASHINGTON   USA
ANDREETTI,G.D., IST.DI CHIM.FISICA, UNIV.OF PARMA, VIA D,AZEGLIO 85 PARMA ITALY
ANZENHOF. SEE ANZENHOFER
ANZENHOFER, K., KONINKLIJKE/SHELL LABORATORIUM, AMSTERDAM, NETHERLANDS
APPEL, K., QUANTUM CHEMISTRY GROUP, UNIV. OF UPPSALA, UPPSALA, SWEDEN
ASBRINK, S., INST. INORG. AND PHYS. CHEM., UNIV. OF STHLM., STOCKHOLM VA, SWEDEN
ASB, SEE ASBRINK, S.
ASHIDA, T., INSTITUTE FOR PROTEIN RES., OSAKA UNIV., KITA-KU, OSAKA, JAPAN
ASHWORTH, E. R., COMP. CENTER, SOUTHERN ILLINOIS UNIV., CARBONDALE, ILL. 62903
BAEHR, S., DAW-STRUKTURFORSCHUNG, BERLIN, GDR
BAENZIGER, N. C., IOWA UNIVERSITY, IOWA CITY, IOWA
BAERNIG, SEE BAERNIGHAUSEN, H.
BAERNIGHAUSEN, H., CHEM. LAB. UNIV. 78 FREIBURG/BRSG., ALBERTSTR. 21, GERMANY
BALLARD,MISS J.V., PHYS.DEPT., COLLEGE OF SCI. AND TECH., MANCHESTER 1, ENGLAND
BARTL,H.,MIN.INST.UNIV. FRANKFURT,SENCKENBERG ANLAGE 30, GERMANY
BASSI, G., CENTRE D*ETUDES NUCLEAIRES DE GRENOBLE, GRENOBLE, ISERE, FRANCE
BAUR, W.H.,DEPT.EARTH+PLANET.SCI.,UNIVERSITY OF PITTSBURGH,PITTSBURGH,PA.15213
BEITINGER, E., MUENCHEN-OTTOBRUNN, FA. BOELKOW, FINKENSTR. 15, GERMANY
BENOFFI ANDREA, OLIVETTI, VIA CLERICI 4, MILANO, ITALY
BENVENUT, SEE BENVENUTI P
```

BENVENUTI P, CSCE, SERVIZIO CALCOLI, LUNGARNO, PACINOTTI 16, PISA, ITALY
BERGERHOFF,G., CHEM. INST. UNIV.,BONN, GERMANY
BERG., SEE BERGERHOFF
BERNARD, W. H., PHYSICS DEPT., LOUISIANA POLYTECHNIC INST., RUSTON, LA., U.S.A.
BERNSTEIN, J.L. BELL TELEPHONE LABORATORIES, MURRAY HILL,NEW JERSEY
BERTOLUZ, SEE BERTOLUZZA C
BERTOLUZZA, C, CSCE, SERVIZIO CALCOLI, LUNGARNO, PACINOTTI 16, PISA, ITALY
BEURSKENS, P.T., LAB. KRISTALCHEMIE, RIJKSUNIVERSITEIT, UTRECHT, NETHERLANDS
BEU, K. E., GOODYEAR ATOMIC, BOX 628, PORTSMOUTH, OHIO
BIBIAN,F,C.E.N.SACLAY,S.P.S.R.M.,BP2,GIF-SUR-YVETTE,78,FRANCE
BIEDL,A.W., LEHRSTUHL FUR KRIST., RUHR-UNIVERSITAET, 463 BOCHUM,GERMANY
BLAND, J. A., I.C.I. LTD., BILLINGHAM DIV., BILLINGHAM CO., DURHAM, ENGLAND
BLOCK, S., NATIONAL BUREAU OF STANDARDS, WASHINGTON DC
BLOMK, SEE BLOMQVIST, G.
BLOMQVIST, G., A.B. ATOMENERGI, STUDSVIK, TYSTBERGA, SWEDEN
BLOUNT, J., DEPT OF CHEM, UNIVERSITY OF WISCONSIN, MADISON, WISCONSIN
BOMBIERI GABRIELLA, CENTRO DI STRUTTUR. CHIM., VIA F.MARZOLO, 1-PADOVA, ITALY
BOMB, SEE BOMBIERI GABRIELLA
BONNER, R. U., SHELL DEVELOPMENT CO., EMERYVILLE, CALIF.
BOOM,G., INSTITUUT VOOR KRISTALFYSICA, RIJKSUNIVERSITEIT, GRONINGEN, NETHERLANDS
BOONSTRA, E. G., N.PHYS.R.L., C.S.I.R., BOX 395, PRETORIA, SOUTH AFRICA
BOOTH, D. P., BELL TELEPHONE LABORATORIES, MURRAY HILL, N J
BORN, L., MIN.-PETRO INST., KIEL, GERMANY
BOWLER,MISS M.,AERE HARWELL, DIDCOT, BERKS, ENGLAND
BOYKO, E. R., WESTINGHOUSE ELECTRIC CO., BETTIS PLANT, PITTSBURGH, PA.
BRADACZEK,H.,FRITZ-HABER INST., ABT. PROF. HOSEMANN, BERLIN 33, GERMANY
BRACHER, B. H., BLDG. 775, AERE, HARWELL, DIDCOT, BERKS, U.K.
BRANDEN, C. I., AGRICULTURAL COLLEGE OF SWEDEN, UPPSALA 7, SWEDEN
BRAUN,R.   UNIVERSITY OF WASHINGTON   SEATTLE, WASHINGTON USA
BROWN, B. W., DEPT. CHEM., PORTLAND STATE COLLEGE, PORTLAND, ORE. 97207
BROWN, G. M., CHEM. DIV., OAK RIDGE NAT. LAB., OAK RIDGE TENNESSEE, U. S. A.
BROWN,B.W.   PORTLAND STATE COLLEGE PORTLAND ORE. USA
BRUHN, RECHENINSTITUT PROF HAACK, TU BERLIN
BRWN, SEE BROWN, B. W.
BRYDEN, J. H., CHEM. DEPT., CALIF. STATE COLLEGE, FULLERTON, CALIF., U. S. A.
BUJOSA, A., I.B.M., S.A.E., SERRANO 5, MADRID 1, SPAIN
BUJ, SEE BUJOSA, A.
BURKE, MARY E., DEPT CHEM, UNIV OF CALIF, LOS ANGELES 24, CALIF
BURNHAM, C. W., GEOPHYSICAL LAB., 2801 UPTON ST., N.W., WASHINGTON 8, D.C.
BUSING,W. R., OAK RIDGE NATIONAL LABORATORY, OAK RIDGE, TENNESSEE
CANUT, M. L., SCHOOL OF TECH., SOUTHERN ILLINOIS UNIV., CARBONDALE, ILL. 62903
CARPENTER, C.L., STATE UNIVERSITY OF IOWA, IOWA CITY, IOWA
CARPENTER, G. B., BROWN UNIVERSITY, PROVIDENCE, RI
CARTER, F. L., WESTINGHOUSE RESEARCH LABORATORIES, PITTSBURGH 35, PA.
CARTER, J. M., GOODYEAR ATOMIC, BOX 628, PORTSMOUTH, OHIO
CETLIN,B.B.,BELL TELEPHONE LABORATORIES,MURRAY HILL,N.J.,USA
CHAND, SEE CHANDROSS, R.
CHANDROSS, R., DEPT. OF CHEM., MIT, CAMBRIDGE, MASS.
CHAST SEE CHASTAIN,R.V.
CHASTAIN,R.V. UNIVERSITY OF WASHINGTON SEATTLE WASH. USA
CHU, S. C. (MRS), CRYSTALLOGRAPHY LAB., UNIVERSITY OF PITTSBURGH,PITTSBURGH,PA.
CLOSS, S. W.,
CLTR, SEE COULTER, C.L.
COHEN,J B ,   DEPT.MATERIALS SC.,TECH INST., NORTHWESTERN UNIV.,EVANSTON,ILL.
COJAZZI GIANNA, CENTRO DI STRUTTUR. CHIM., VIA F.MARZOLO, 1-PADOVA, ITALY
COJAZ, SEE COJAZZI GIANNA
COLASANTI,A.,ISTITUTO CHIMICO,VIA MEZZOCANNONE 4,NAPOLI,ITALIA
COL,SEE COLASANTI,A.
COONEY, W. A., SEE HAENDLER, H. M.
CORFIELD, P., CRYSTALLOGRAPHIC LABORATORY, U. OF PITTSBURGH, PITTSBURGH 13, PA.
COULTER, DR. C.L. M.R.C. UNIT, CAVENDISH LABORATORY, CAMBRIDGE, ENGLAND
CRUICK, SEE CRUICKSHANK, D.W.J.
CRUICKSHANK, D. W. J., CHEM. DEPT., THE UNIVERSITY, GLASGOW, W.2, SCOTLAND
CURTIS,A.R.,AERE HARWELL, DIDCOT, BERKS, ENGLAND
DAHL, L.F., DEPT OF CHEM, UNIVERSITY OF WISCONSIN, MADISON, WISCONSIN
DALLINGA, G., KONINKLIJKE/SHELL LABORATORIUM, AMSTERDAM, NETHERLANDS

DAMIANI,A.,ISTITUTO CHIMICO,VIA MEZZOCANNONE 4,NAPOLI,ITALIA
DANIELSEN, J., DEP. OF INORG. CHEM., AARHUS UNIVERSITY, AARHUS, DENMARK
DANIELS, SEE DANIELSEN, J.
DARLOW, S.F., PHYS. DEPT., COLLEGE OF SCI. AND TECH., MANCHESTER 1, ENGLAND
DARLOW, MRS J. V., SEE DARLOW, S. F.
DAYHOFF, M. O.,NAT'L BIOMEDICAL RESEARCH FDN. 8600 16 ST. SILVER SPRING, MD.
DAYH, SEE DAYHOFF, M. O.
DEANGELIS, R. J., DEPT. OF METALL. ENG., U. OF KENTUCKY, LEXINGTON, KY.
DEGRIF, SEE DE GRIFFI ELENA
DE GRIFFI ELENA, OLIVETTI, VIA CLERICI, MILANO, ITALY
DELF, B. W., PHYS. DEPT., UNIVERSITY COLLEGE, CARDIFF, WALES
DENNER,K H ,DAW,INSTITUT FUER STRUKTURFORSCHUNG,BERLIN-ADLERSHOF,GDR
DE VRIES, A., LAB. KRISTALCHEMIE, RIJKSUNIVERSITEIT, UTRECHT, NETHERLANDS
DIA, SEE DIAMAND, R. D.
DIAMAND, R. D., ROYAL INSTITUTION, LONDON W.1, ENGLAMD
DICKINSON,C.W. NAVAL ORDNANCE LAB. WHITE OAK, SILVER SPRING, MD. USA
DICSN SEE DICKINSON,C.W.
DIDRIKSEN, F., IBM, OSLO, NORWAY
DIETRICH,H., FRITZ'HABER'INSTITUT,BERLIN'DAHLEM,FARADAYWEG 4'6, GERMANY
DODGE, R. UNION CARBIDE RES. INST., BOX 324, TUXEDO, N.Y.
DOL, SEE DOLLIMORE, J.
DOLLIMORE,J., UNIV.LOND.INST.COMPUTER SCI., 44 GORDON SQ., LONDON WC1, ENGLAND
DOMENICANO, A., CENTRO STRUTTURISTICA CNR, CITTA UNIVERSITARIA, ROMA, ITALY
DOMENICI,M., SORIN, SALUGGIA, ITALY
DOME, SEE DOMENICANO, A.
DOMIANO,P., CENTRO DI CALC.ELETTR., UNIV.OF PARMA,VIA M.D.AZEGLIO 85 PARMA ITALY
DONNAY, J.D.H., CRYSTAL. LAB., JOHNS HOPKINS UNIV., BALTIMORE, MD.21218
DREW, M. G. B., CHEM. DEPT., IMPERIAL COLLEGE, LONDON S.W.7, ENGLAMD
DUCHAMP,D.J.,THE UPJOHN COMPANY,KALAMAZOO,MICHIGAN
DUISENBERG, A.J.M., LAB.KRISTALCHEMIE, RIJKSUNIVERSITEIT, UTRECHT, NETHERLANDS
EDSTRAND, M., FORSKN. AVD., A.B. BOFORS NOBELKRUT, BOFORS, SWEDEN
EHRLICH, H. W. W., ICI LTD, HEAVY ORG. CHEMICALS DIV., BILLINGHAM, CO DURHAM, UK
EICHHORN, E. L., 7802 ALLAN STRUGES TERR., FALLS CHURCH, VA. 22041, USA
ELLISON, R.D., OAK RIDGE NATIONAL LABORATORY, OAK RIDGE, TENNESSEE.
ESPOSITO,U.,ISTITUTO CHIMICO,VIA MEZZOCANNONE 4,NAPOLI,ITALIA
ESP,SEE ESPOSITO,U.
ESQUIVEL,A.L.,MAT LS RES LAB,MP-105,MARTIN CO.,ORLANDO,FLA   32805
FARRAR, R. A., FACULTY OF ENG., KWAME NKRUMA UNIV., KUMASI, GHANA
FENN, R.H., PHYS.DEPT., PORTSMOUTH COLL.OF TECH., PARK RD.,PORTSMOUTH,HANTS, UK
FITZW, SEE FITZWATER, D.
FITZWATER, D., IOWA STATE UNIVERSITY, AMES, IOWA
FRASS, SEE FRASSON EDOARDO
FRASSON EDOARDO, CENTRO DI STRUTTURISTICA CHIMICA, VIA F.MARZOLO,1, PADOVA,ITALY
FREEMAN, H. C., SCH. OF CHEM., UNIVERSITY OF SYDNEY, SYDNEY, AUSTRALIA
FREER,S.   UNIVERSITY OF CALIFORNIA AT SAN DIEGO LA JOLLA CALIF. USA
GABE, E.J., INST. FOR CANCER RESEARCH, 7701 BURHOLME AVE., PHILADELPHIA 11, PA.
GALLAHER,L.J. E.E.S. GEORGIA INST. OF TECHNOLOGY,ATLANTA,GA. U.S.A.
GANTZEL, PETER K. , DEPT CHEM, UNIV OF CALIF, LOS ANGELES 24, CALIF
GEISE, H. J., LAB. ORGANIC CHEMISTRY, UNIV. LEIDEN, LEIDEN, THE NETHERLANDS
GERGELY, G., COMPUTING CENTER OF THE HUNG. ACAD. OF SCIENCES, BUDAPEST, HUNGARY
GIGLIO,E.,ISTITUTO CHIMICO,VIA MEZZOCANNONE 4,NAPOLI,ITALIA
GIG,SEE GIGLIO,E.
GILMARTIN,MRS J.,AERE HARWELL, DIDCOT, BERKS, ENGLAND
GOEBEL,J.B., APPLIED MATH., BATTELLE-NORTHWEST, RICHLAND, WASHINGTON
GRAF, D. L., ILLINOIS GEOLOGICAL SURVEY, URBANA, ILL.
GRANT, D. F., PHYS. DEPT., UNIVERSITY COLLEGE, CARDIFF, WALES
GREEN, D.W. ROYAL INSTITUTION, LONDON W.1, ENGLAND
GROENBAEK, RITA, DEPT. OF INORG. CHEM., AARHUS UNIVERSITY, AARHUS, DENMARK
GRUBISS, SEE GRUBISSICH CLAUDIO
GRUBISSICH CLAUDIO, C.E.C.S., VIA PAOLOTTI 9, PADOVA, ITALY
GTZL, SEE GANTZEL, PETER K.
GUERRI L., CSCE, SERVICIO CALCOLI, LUNGARNO, PACINOTTI 16, PISA, ITALY
HAENDLER, H. M., DEPT CHEM, UNIV OF NH, DURHAM, NH
HAHN,TH.,INST.FUR KRIST.,TECHN.HOCHSCHULE,51 AACHEN,TEMPLERGRABEN 55, GERMANY
HALL,S.R.,PHYSICS DEPT.,UNIV. OF W.A.,NEDLANDS,WESTERN AUSTRALIA
HAMILTON, W. C., BROOKHAVEN NATIONAL LABORATORY, BROOKHAVEN, NY

HAMOR, T. A., CHEM. DEPT., UNIV. OF BIRMINGHAM, BIRMINGHAM 15, ENGLAND
HARD, SEE HARDING, M. M.
HARDING, MRS. M. M., CHEM. CRYST. LAB., SOUTH PARKS ROAD, OXFORD, ENGLAND
HARDING,M.M., EDINBURGH UNIV. CHEM. DEPT., WEST MAINS ROAD, EDINBURGH 9,SCOTLAND
HARRIS,D.R. DEPT. OF COMP. SCIENCE, UTAH STATE UNIV., LOGAN, UTAH
HEATON, L. ARGONNE NATIONAL LABORATORY, ARGONNE, ILLINOIS
HECKEL, R.W., DEPT. OF MET. ENG., DREXEL INST. OF TECH., PHILADELPHIA 4, PA.
HELLNER, E., MIN.-PETRO. INST., KIEL, GERMANY
HENDE, SEE VAN DEN HENDE, J. H.
HESPER, B., LAB. BIOCHEMISTRY, UNIV. LEIDEN, LEIDEN, THE NETHERLANDS
HIGH,D.F. UNIVERSITY OF CALIFORNIA AT SAN DIEGO LA JOLLA CALIF. USA
HILDEB, SEE HILDEBRAND
HILDEBRAND, R., MIN. INST. UNIV., FRANKFURT , SENCKENBERG ANLAGE 30,GERMANY
HINE. R., PHYSICS DEPT., UNIVERSITY COLLEGE, CARDIFF, WALES
HIRSHFELD, F., DEPT. OF CHEM, UNIV. OF MINN., MINNEAPOLIS, MINN.
HOLDEN,J.R. NAVAL ORDNANCE LAB. WHITE OAK, SILVER SPRING, MD. USA
HOP., SEE HOPPE
HOPPE,W., MPI. EIWEISSFORSCHUNG, MUENCHEN, LUISENSTR. 39, GERMANY
HOUSTON, B. J.,
HUBER,R., MPI. EIWEISSFORSCHUNG, MUENCHEN, LUISENSTR. 39, GERMANY
HUNT,   , CHEM. DEPT., IMPERIAL COLLEGE, LONDON S.W. 7, ENGLAND
IANDOLO,A.,ISTITUTO CHIMICO,VIA MEZZOCANNONE 4,NAPOLI,ITALIA
IAN,SEE IANDOLO,A.
IBERS, J. A., DEPT. CHEM., NORTHWESTERN UNIV., EVANSTON, ILL., USA
IITAKA, Y., FAC. OF PHARMACEUTICAL SCI., UNIV. OF TOKYO, BUNKYO-KU, TOKYO, JAPAN
JACOBSON, R. A., DEPT. OF CHEM., UNIV. OF MINN., MINNEAPOLIS, MINN.
JAMARD,C., 91 RUE VILLIERS DE L,ISLE ADAM, PARIS 20E, FRANCE
JAMES,   , BLDG. 775, AERE, HARWELL, DIDCOT, BERKS., U.K.
JEFFREY, G. A., CRYSTALLOGRAPHIC LAB., UNIVERSITY OF PITTSBURGH, PITTSBURGH, PA.
JENSEN, L. H., DEPT. BIOL. STRUCT., UNIV. OF WASHINGTON, SEATTLE, WASH. 98105
JMS, SEE JAMES
JOHANSSON, G., DEPT. OF INORG. CHEM., INST. OF TECHNOLOGY, STOCKHOLM 70, SWEDEN
JOHNSON, C. K., OAK RIDGE NATIONAL LABORATORY, OAK RIDGE, TENNESSEE, U.S.A.
JOHNSON, Q., UNIV. OF CALIFORNIA, BERKELEY, CALIF.
JONES,   UNIV. OF CALIFORNIA, BERKELEY, CALIF.
JSTEW SEE STEWART,J.M.
KAPLOW, R., METALLURGY DEPT., MIT, CAMBRIDGE, MASS.
KASPER, H., CHEM. INST. UNIV. 53 BONN, MECKENHEIMER ALLEE 168, GERMANY
KATZ, L., DEPT. OF BIOLOGY, M.I.T., CAMBRIDGE MASS. 02139, U.S.A.
KAY,M.I. E.E.S. GEORGIA INST. OF TECHNOLOGY, ATLANTA, GA. U.S.A.
KEEFE,W. MEDICAL COLLEGE OF VIRGINIA RICHMOND, VA. USA
KEILHAU, H., NORWEGIAN DEFENCE RESEARCH ESTABLISHMENT, KJELLER, NORWAY
KEMPTER,C.P.,LOS ALAMOS SCIENTIFIC LABORATORY,LOS ALAMOS,NM
KEULEMANS, E. W. M., LAB. VOOR KRISTALLOGRAFIE, UNIV. OF AMSTERDAM, NETHERLANDS
KEUL, SEE KEULEMANS, E. W. M.
KEUNING, W., CENTR. PROEFSTATION, STAATSMIJNEN, HOENSBROEK, NETHERLANDS
KEYES, P. A., IOWA STATE UNIVERSITY, AMES, IOWA
KING, G. S. D., E. R. A., 95 RUE GATTI DE GAMOND, BRUSSELS 18, BELGIUM
KOENE, A. A., STATISTICAL DEPT., T.N.O., THE HAGUE, NETHERLANDS
KOENIG, D. F., BIOLOGY DEPT., BROOKHAVEN NAT. LAB., UPTON, LONG ISLAND, N.Y.,USA
KRAUSE,CH ,DAW,INSTITUT FUER STRUKTURFORSCHUNG,BERLIN-ADLERSHOF,GDR
KRAUT, J., DEPT.CHEM., U. OF CAL.SAN DIEGO., LA JOLLA, CALIF. 92038, USA
KRETSCHMER,R G ,DAW,INSTITUT FUER STRUKTURFORSCHUNG,BERLIN-ADLERSHOF,GDR
KRUECKEBERG,F., MATH. INST. UNIV. BONN,GERMANY
KRUECKEBG, SEE KRUECKEBERG
LAGERWALL, X., ADP INST., CHALMERS UNIV. OF TECHN., GIBR.G.5, GOTHENBURG, SWEDEN
LAI, T. F., CHEM. DEPT., UNIVERSITY OF HONG KONG
LANGHAMMER, D., PHYSICS DEPT., BRADLEY UNIV., PEORIA, ILLINOIS, U. S. A.
LANGHMR, SEE LANGHAMMER, D.
LARSEN, F., DANISH INST. FOR COMPUTING MACHINERY, GL CARLSBERGV 2, VALBY,DENMARK
LARSON, A.C. LOS ALAMOS NATL. LAB., LOS ALAMOS, N.M.
LEDLEY, R. S., NAT'L BIOMEDICAL RESEARCH FDN. 8600 16 ST. SILVER SPRING, MD.
LEFKER, R., USASRDL-XE, FORT MONMOUTH, NJ
LEMM,K., FRITZ-HABER-INST. M.P.G., ABT. PROF. HOSEMANN, BERLIN-33, GERMANY
LENHERT, P.GALEN., PHYS. DEPT., VANDERBILT UNIV., NASHVILLE, TENN.
LESJAK, J., COMPUTING CENTRE, LEPI POT 11, LJUBLJANA, JUGOSLAVIA

LEVY, H. A., OAK RIDGE NATIONAL LABORATORY, OAK RIDGE, TENNESSEE
LIMINGA, R., DEPT. OF CHEMISTRY, UNIV. OF UPPSALA, UPPSALA, SWEDEN
LIND, M.D., UNION OIL RESEARCH CTR., BREA, CALIF.92621, USA
LINEK, A., INST. OF SOLID STATE PHYS., CSAV, PRAGUE, CZECHOSLOVAKIA
LING   SEE LINGAFELTER,E.C.
LINGAFELTER, E. C., DEPT. CHEM., UNIV. OF WASHINGTON, SEATTLE, WASH. 98105
LINGFTR, SEE LINGAFELTER, E. C.
LIPSCOMB, W. N., JR., DEPT. OF CHEM., HARVARD UNIVERSITY, CAMBRIDGE, MASS.
LIPSC, SEE LIPSCOMB, W. N., JR.
LOMBARD, SEE LOMBARDI SILVANA
LOMBARDI SILVANA, OLIVETTI, VIA CLERICI, MILANO, ITALY
LOVELL, F.M. SCHOOL OF CHEM., UNIV. OF SYDNEY, N.S.W., AUSTRALIA
LOVE, W. E., BIOPHYSICS DEPT., JOHNS HOPKINS UNIV., BALTIMORE, MD.
LOV, SEE LOVELL, F.M.
LUNDBERG, B.K.S., UNIVERSITY OF UMEA, UMEA, SWEDEN
LUNDBRG, SEE LUNDBERG, B.
MAC SEE MACINTYRE, W.M.
MACGILL, SEE MACGILLAVRY, C. H.
MACGILLAVRY, C. H., LAB. VOOR KRISTALLOGRAFIE, UNIV. OF AMSTERDAM, NETHERLANDS
MACINTYRE, W. M., DEPT. OF CHEM., UNIVERSITY OF COLORADO, BOULDER, COLO.
MAIN, P., PHYSICS DEPT., COLLEGE OF SCI. AND TECHNOLOGY, MANCHESTER 1, ENGLAND
MAIR, G.A., ROYAL INSTITUTION, 21 ALBERMARLE ST., LONDON W.1, ENGLAND
MALLETT, G. R., DOW CHEMICAL CO., BOX 2131, DENVER, COLO.
MARIANI C., INSTITUTO DO CHIMICA FISICA, UNIVERSITA DI MILANO, MILANO, ITALY
MARSH, R. E., CALIF. INST. OF TECH., PASADENA, CALIF
MARTIN,K. O., OAK RIDGE NATIONAL LABORATORY, OAK RIDGE, TENNESSEE
MAR, SEE MARTIN,K. O.
MASAKI, N., FAC. OF PHARMACEUTICAL SCI., KYOTO UNIV., SAKYO-KU, KYOTO, JAPAN
MASLEN,E.N.,PHYSICS DEPT.,UNIV. OF W.A.,NEDLANDS,WESTERN AUSTRALIA
MATHER, K., P. O. DR. 2131 JACKSON, MISS.
MATHEWS,
MATTES,R., LABORATORIUM FUER ANORG. CHEMIE DER T.H. STUTTGART, GERMANY
MCGANDY, E. L., CHEM. DEPT., BOSTON UNIVERSITY, BOSTON, MASS.
MCGREGOR,D.R., CHEM. DEPT., THE UNIVERSITY, GLASGOW W.2, SCOTLAND
MCMULLAN, R. K., DEPT. OF CHEM., UNIV. OF PITTSBURGH, PITTSBURGH, PENNSYLVANIA
MEDRUD, R. C., IOWA UNIVERSITY, IOWA CITY, IOWA
MEYER, E.F.H., DEPT. CHEMISTRY, TEXAS UNIVERSITY, AUSTIN, TEXAS
MIGLIORISI, G., SORIN, SALUGGIA, ITALY
MIGLIO, SEE MIGLIORISI, G.
MILLER, D. P., TEXAS INSTITUTE, DALLAS, TEXAS
MINKIN,J.A., INST. FOR CANCER RESEARCH, 7701 BURHOLME AVE.,PHILADELPHIA PA.19111
MOHN, G. J., WESTINGHOUSE ELECTRIC CO., BETTIS PLANT, PITTSBURGH, PA.
MONDRUP, P.,DANISH INST. FOR COMPUTING MACHINERY, GL CARLSBERGV 2,VALBY,DENMARK
MOORE, F. H., BLDG. 775, AERE, HARWELL, DIDCOT, BERKS., U.K.
MORE, SEE MOORE, F. H.
MORGAN, C. H., CHEM. DEPT., QUEENS COLLEGE, DUNDEE, SCOTLAND
MORINO, Y., DEPT. OF CHEM., UNIV. OF TOKYO, BUNKYO-KU, TOKYO, JAPAN
MOROSIN,B. SANDIA CORPORATION ALBUQUERQUE, NEW MEXICO USA
MORROW, J. C., UNIV. OF NORTH CAROLINA, CHAPEL HILL, NC
MOSELEY,JERRY, CHEMISTRY DEPT., UNIVERSITY OF COLORADO,BOULDER,COLO.
MOZZI, R., RESEARCH DIV., RAYTHEON CO., WALTHAM 54, MASS.
MUELLER, M.H.   ARGONNE NATIONAL LABORATORY ARGONNE, ILLINOIS
MUIR,K.W., CHEM. DEPT., THE UNIVERSITY, GLASGOW W.2, SCOTLAND
MUSATTI,A., CENTRO DI CALC.ELETTR., UNIV.OF PARMA,VIA M.D,AZEGLIO 85 PARMA ITALY
MUSIL, F. J., GOODYEAR ATOMIC, BOX 628, PORTSMOUTH, OHIO
NADRCHAL, J., INST. OF SOLID STATE PHYS., CSAV, PRAGUE, CZECHOSLOVAKIA
NADR, SEE NADRCHAL, J.
NEETHLING, J. D., N. P. R. L., PRETORIA, SOUTH AFRICA
NEWELL, J., RESEARCH DIV., RAYTHEON CO., WALTHAM 54, MASS.
NIIZEKI, N., ELEC. COMM. LAB. JAPAN TEL-TEL PUB. CORP., TOKYO, JAPAN
NKB, SEE NOVAK, B.
NORMENT, H., CALLERY CHEMICALS,
NORRESTAM,R., INST. INORG. AND PHYS. CHEM., UNIV. OF STHLM., STOCKHOLM VA,SWEDEN
NORTH, A.C.T., ROYAL INSTITUTION, 21 ALBEMARLE ST., LONDON W. 1, ENGLAND
NOVAK, B., NUM. CENTR. OF MAT., CHARLES UNIV., PRAGUE, CZECHOSLOVAKIA
NOVAK, C., INST. OF SOLID STATE PHYS., CSAV, PRAGUE, CZECHOSLOVAKIA

NVK, SEE NOVAK, C.
NYBORG, J., DEP.OF INORG. CHEM., AARHUS UNIVERSITY, AARHUS, DENMARK.
OCENASKOVA,D., INST . OF SOLID STATE PHYS., C.S.A.V., PRAGUE, CZECHOSLOVAKIA
OCKEN, H. HAMMOND LABORATORY, YALE UNIVERSITY, NEW HAVEN, CONN.
OCONNOR,B.H.,PHYSICS DEPT.,UNIV.OF W.A.,NEDLANDS,WESTERN AUSTRALIA
OGAWA, Y., INSTITUTE FOR PROTEIN RESEARCH, OSAKA UNIV., KITAKU OSAKA JAPAN
OLOVSON, SEE OLOVSSON, I.
OLOVSSON, I., DEPT. OF CHEMISTRY, UNIV. OF UPPSALA, UPPSALA, SWEDEN
ONKEN, H., LEHRST. F. KRISTALLOGR., UNIV. SAARBRUECKEN, WEST GERMANY
OSAKI, K., FAC. OF PHARMACEUTICAL SCI., KYOTO UNIV., SAKYO-KU, KYOTO, JAPAN
OTTE,H.M., MAT LS RES LAB,MP-105,MARTIN CO,ORLANDO,FLA.   32805
PALENIK,G.J.,CODE 5058, USNOTS ,CHINA LAKE,CALIFORNIA
PALMER,R.A.,BIRKBECK COLLEGE RES.LAB., 21 TORRINGTON SQ., LONDON W.C.1, ENGLAND
PALM, J. H., LAB. TECHN. PHYSICA, TECHNISCHE HOGESCHOOL, DELFT, NETHERLANDS
PANATT, SEE PANATTONI CARLO
PANATTONI CARLO, CENTRO DI STRUTTURISTICA CHIMICA, VIA F.MARZOLO,1,PADOVA,ITALY
PATT SEE PATTERSON, A.L.
PATTERSON, A.L., INST FOR CANCER RESEARCH, 7701 BURHOLME AVE, PHILADELPHIA 11,PA
PAULING, PETER, CHEM. DEPT., UNIVERSITY COLLEGE, GOWER ST., LONDON W.C.1,ENGLAND
PEACOR, D.R., DEPT OF GEOL AND GEOPHYS, MIT, CAMBRIDGE 39, MASS
PENFOLD, B., DEPT. OF CHEM., HARVARD UNIVERSITY, CAMBRIDGE, MASS.
PEPINSKY, R., DEPT. OF PHYSICS, PENN. STATE UNIV., UNIVERSITY PARK, PA.
PETERS, MRS. G., MATH. DIV., NATIONAL PHYSICAL LAB., TEDDINGTON, MIDDX., ENGLAND
PETERSON, D. P., DOW CHEMICAL COMPANY, MIDLAND, MICH.
PETERSE,W.J.A.M.,LAB.TECHN.PHYSICA,TECHNISCHE HOGESCHOOL,DELFT,NETHERLANDS
PIPPY, M.E., NATIONAL RESEARCH COUNCIL, OTTAWA 2, ONTARIO, CANADA
PISTORIUS, C. W. F. T., N.PHYS.R.L., C.S.I.R., BOX 395, PRETORIA, SOUTH AFRICA
PISTORIUS, M.C., N.R.I.MATH.SC., C.S.I.R., BOX 395, PRETORIA, SOUTH AFRICA
PLEHWE, A. VON, INST. ANG. MATH. UNIV. 78 FREIBURG/BRSG., HEBELSTR. 40, GERMANY
PREWITT, C. T., DEPT. OF GEOLOGY, MASS. INST. OF TECH., CAMBRIDGE, MASS.
PROUT, C. K., CHEM. CRYST. LAB., SOUTH PARKS RD., OXFORD, ENGLAND
PTRSON, SEE PETERSON, D. P.
RAE,A.I.M.,PHYSICS DEPARTMENT,UNIVERSITY OF W.A.,NEDLANDS,WESTERN AUSTRALIA.
RAE,A.I.M.,PHYSICS DEPT.,UNIV.OF W.A.,NEDLANDS,WESTERN AUSTRALIA
RICHARDS, J. P. G., PHYS. DEPT., UNIVERSITY COLLEGE, CARDIFF, WALES
ROGERS, D., CHEMISTRY DEPT., IMPERIAL COLLEGE, LONDON, S.W.7, ENGLAND
ROLLETT, J. S., OXFORD UNIV. COMPUTING LAB., SOUTH PARKS ROAD, OXFORD, ENGLAND
ROMMING, CHR., CHEMISTRY DEPARTMENT, UNIVERSITY OF OSLO, BLINDERN, NORWAY
ROSSMANN,M.G.,M.R.C.LAB OF MOLECULAR BIOLOGY,HILLS RD.,CAMBRIDGE,ENGLAND
ROWE,J.M., I.C.I.LTD.,HEAVY ORG. CHEMICALS DIV., THE FRYTHE,WELWYN,HERTS,ENGLAND
RUDEE, M.L., DEPT. OF MATERIALS SCI., STANFORD, CALIFORNIA
RUDOLPH, G. J., N.R.I.MATH.SC., C.S.I.R., BOX 395, PRETORIA, SOUTH AFRICA
RUTTEN-KEULEMANS, E. W. M., CENTRAAL REKENINST., UNIV. LEIDEN, THE NETHERLANDS
RUTTEN-KEULEMAN, SEE RUTTEN-KEULEMANS, E. W. M.
SAMET P., CO H.P. STADLER, KINGS COLLEGE, NEWCASTLE UPON TYNE, ENGLAND
SANBORN, M. A., I.B.M. MFG. RES. LAB., ENDICOTT, N. Y.
SANDS, D. E., LAWRENCE LABORATORY, LIVERMORE, CALIF.
SANTA, L., COMPUTING CENTER OF THE HUNG. ACAD. OF SCIENCES, BUDAPEST, HUNGARY
SASADA, Y., INSTITUTE FOR PROTEIN RESEARCH, OSAKA UNIV., KITA-KU, OSAKA, JAPAN
SASS, R., RICE INSTITUTE, HOUSTON, TEXAS
SASVARI, K., CRYST. LAB. U. OF PITT., PITTSBURGH 13, PA, ON LEAVE FROM HUNGARY
SAYRE, D., IBM CO., 590 MADISON AVE., NYC
SCHAPIRO, P. J., DEPT. OF CHEM., U. OF OKLAHOMA, NORMAN, OKLAHOMA, USA
SCHMDMT, SEE SCHMITZ-DU MONT, O
SCHMITZ-DU MONT, O., CHEM. INST. UNIV. 53 BONN, MECKENHEIMER ALLEE 168, GERMANY
SCHMID,E., RECHENINSTITUT DER TECHN. HOCHSCHULE, STUTTGART, HERDWEG 51 GERMANY
SCHOONE, J. C., LAB. VOOR KRISTALCHEMIE, UNIVERSITY OF UTRECHT, NETHERLANDS
SCHULZ, H., MINERAL. INST. D. UNIVERS., HAMBURG 13, GRINDELALLEE 48, GERMANY
SCHULTZERHONHOF, SEE SCHULTZE-RHONHOF,E.
SCHULTZE-RHONHOF, E., CHEM. INST. UNIV. 53 BONN, MECKENHEIMER ALLEE 168, GERMANY
SCHWARTZ,L.H.,MAT. SCI. DEPT., NORTHWESTERN U. EVANSTON,ILL.
SEGMUELLER, A., IBM RESEARCH LABORATORY, 8803-RUESCHLIKON, SWITZERLAND
SENKO, M. E., IBM COMPANY, POUGHKEEPSIE, N Y
SHEFTER,   , BLDG. 775, AERE, HARWELL, DIDCOT, BERKS., U.K.
SHFTER, SEE SHEFTER
SHIONO, R., CRYSTALLOGRAPHIC LAB., UNIVERSITY OF PITTSBURGH, PITTSBURGH. PA.

SHOEMAKER, C. B., DEPT. OF CHEM., MIT, CAMBRIDGE, MASS.
SHOEMAKER, D. P., DEPT. OF CHEM., MIT., CAMBRIDGE, MASS.
SHO, SEE SHOEMAKER, D. P.
SILVERTON, J. V., CORNELL UNIV., DEPT. OF CHEMISTRY, ITHACA, NY.
SIME, J.G., CHEM. DEPT., THE UNIVERSITY, GLASGOW W.2, SCOTLAND
SIMPSON, P. G., DEPT. OF CHEM., STANFORD UNIV., STANFORD, CALIF. 94305, USA
SIMPSON, S.M., DEPT. OF GEOL., M. I. T., CAMBRIDGE, MASS. 02139, USA
SLITER, J. A., 2162 14TH ST., TROY, N Y
SMITH, D. K., LAWRENCE RADIATION LABORATORY, LIVERMORE, CALIF.
SMITH, J. G., CHEM. DEPT., THE UNIVERSITY, GLASGOW W2, SCOTLAND
SMITH, A. E., SHELL DEVELOPMENT CO., EMERYVILLE, CALIF.
SMITH, J., SEE SMITH, J.G.F.
SMITH, J. G. F., CHEM. DEPT., THE UNIVERSITY, GLASGOW, W.2, SCOTLAND
SMITS, D. W., REKENCENTRUM DER RIJKSUNIVERSITEIT, GRONINGEN, THE NETHERLANDS
SPARKS, R. A., IBM CORP., TIME-LIFE BLDG., NYC
SPEAKMAN, J. C., CHEM. DEPT., THE UNIVERSITY, GLASGOW, W.2, SCOTLAND
SPKS, SEE SPARKS, ROBERT A.
SRIVASTAVA, R. C., DEPT. OF CHEM., U. OF WASHINGTON, SEATTLE, WASHINGTON
STADLER, H. P., THE UNIVERSITY, NEWCASTLE-UPON-TYNE, ENGLAND
STALEC, F., COMPUTING CENTRE, LEPI POT 11, LJUBLJANA, JUGOSLAVIA
STEMMLER, R. S., ILLINOIS GEOLOGICAL SURVEY, URBANA, ILL.
STEWART, J. M., DEPT. CHEM., UNIV. OF MARYLAND, COLLEGE PARK, MARYLAND
STOUT, G. H., DEPT CHEM, UNIV. OF WASHINGTON, SEATTLE 5, WASH.
STWRT, SEE STEWART, J. M.
SUMNER, G. G., MELLON INST., PITTSBURGH, PA.
SVETICH, G., 7126 S 130TH ST., SEATTLE, WASH., USA
TAKEDA, HIROSHI, CRYSTAL. LAB., JOHNS HOPKINS UNIV., BALTIMORE, MD.21218
TAKEUCHI, Y., DEPT. OF MINERALOGY, UNIV. OF TOKYO, HONGO, TOKYO, JAPAN
TAYLOR, F. C., EES, GEORGIA INSTITUTE OF TECHNOLOGY, ATLANTA, GEORGIA
TEMPLETON, D. H., UNIV, OF CALIFORNIA, BERKELEY, CALIF.
TEMP, SEE TEMPLETON, D. H.
THURN,H.,LEHRST.ANORG.ANALYT.CHEMIE T.H.STUTTGART,SCHELLINGSTR.26,GERMANY
TICHY, K., INST. OF SOLID STATE PHYS., C.S.A.V., PRAGUE, CZECHOSLOVAKIA
TOEPFER, RECHENINSTITUT PROF HAACK, TU BERLIN
TOMAN, K., INST. OF SOLID STATE PHYS., CSAV, PRAGUE, CZECHOSLOVAKIA
TOURNARIE,M.,40,RUE-MANGEON,MASSY,78,FRANCE
TRAUB, W., COLUMBIA UNIVERSITY (COLLEGE OF PHYSICIANS AND SURGEONS), NYC
TRBD, SEE TRUEBLOOD, K. N.
TREUTING, R. G., BROOKHAVEN NATIONAL LABORATORY, BROOKHAVEN, NY
TREUTNG, SEE TREUTING, R. G.
TROTTER, J., DEPT. OF CHEM., UNIV. OF BRIT. COLUMBIA, VANCOUVER, CANADA
TRUEBLOOD, K.N. , DEPT CHEM, UNIV OF CALIF, LOS ANGELES 24, CALIF
TRUTER, M.R., SCHOOL OF CHEM., UNIVERSITY, LEEDS 2, ENGLAND
TULINSKY, A., DEPT. OF CHEMISTRY., YALE UNIVERSITY, NEW HAVEN, CONN.
TUVLIND, S.O., ADP INST., CHALMERS UNIV. OF TECHN., GIBR.G.5, GOTHENBURG, SWEDEN
UEDA,I., GENERAL EDUCATION DEPT., KYUSHU UNIV., OTSUBO-MACHI, FUKUOKA, JAPAN
URBAN, J., FRITZ-HABER-INST., ABT. PROF. HOSEMANN, BERLIN 33, GERMANY
VACIAGO, A., CENTRO STRUTTURISTICA CNR, CITTA UNIVERSITARIA, ROMA, ITALY
VAC, SEE VACIAGO, A.
VAND, V., GROTH INST., PENN. STATE UNIV., UNIVERSITY PARK, PA.
VAN DEN HENDE, J. H., LEDERLE LABAROTORIES, AMER. CYANAMIDE, PEARL RIVER, N. Y.
VAN DER HELM, D. CHEM.DEPT.,UNIV. OF OKLAHOMA, NORMAN, OKLA., USA.
VAN DER SLUIS,   , 51 CATHARIJNESINGEL, UTRECHT, NETHERLANDS
VAN EIJCK, B.P., LAB. KRISTALCHEMIE, RIJKSUNIVERSITEIT, UTRECHT, NETHERLANDS
VDHELM SEE VAN DER HELM, D.
VISSER,J.W.,TECHN,PHYS.DEPT.,P.O.BOX 155, DELFT,NETHERLANDS
VOGEL,R.E., KAMAN CORP COLO SPRINGS,COLO
VON STEINWEHR,H.E., INST.MIN.PETR., UNIVERSITAT, PPOSTFACH 606, 65 MAINZ,GERMANY
VONK, C. G., CENTRAAL LAB. STAATSMIJNEN, GELEEN, NETHERLANDS
VOS,A,LAB. VOOR STRUCTUURCHEMIE BLOEMSINGEL 10, GRONINGEN,NETHERLANDS
WAX,S. NASA GODDARD SPACE FLIGHT CENTER GREENBELT MD USA
WEHE,D. J., OAK RIDGE NATIONAL LABORATORY, OAK RIDGE, TENNESSEE
WEH, SEE WEHE,D. J.
WEISS, H. G., DAW-STRUKTURFORSCHUNG, BERLIN, GDR
WEISS,H G ,DAW,INSTITUT FUER STRUKTURFORSCHUNG,BERLIN-ADLERSHOF,GDR
WENGELIN,R.F.,INORG.CHEM.,UNIV.OF GOTHENBURG,GOTEBORG S,SWEDEN

WERKEMA,MARILYN, CHEMISTRY DEPT., UNIVERSITY OF COLORADO, BOULDER,COLO.
WERNER, P.E., INST. INORG. AND PHYS. CHM., UNIV. OF STHLM., STOCKHOLM VA, SWEDEN
WESTM, SEE WESTMAN, S.
WESTMAN, S., INST. INORG. CHEM., UNIV. OF STOCKHOLM, STOCKHOLM 50, SWEDEN
WHITTIER, V. E., DOW CHEMICAL COMPANY, MIDLAND, MICH.
WILLIAMS, D. E., IOWA STATE UNIVERSITY, AMES, IOWA
WILL,G., MPI. EIWEISSFORSCHUNG,MUENCHEN,LUISENSTR. 39, GERMANY
WILSON, A.S., CHEM. RESEARCH, BATTELLE-NORTHWEST, RICHLAND, WASHINGTON
WOLFEL, E. R., ABT. FUR. STRUKTURFORSCHUNG, DARMSTADT, GERMANY
WOLTEN, G.M. AEROSPACE CORP., BOX 95085, LOS ANGELES 45, CALIF.
WOOLFSON,M.M., MANCHESTER COLLEGE OF SCIENCE AND TECHNOLOGY, MANCHESTER, ENGLAND
WRIGHT, D.A., DEFENSE STANDARDS LABS, BOX 50, ASCOT VALE, W2, VICTORIA, AUSTRAL.
WUENSCH, B. J., DEPT. OF METALLURGY, M. I. T., CAMBRIDGE, MASS., 02139, USA
WUNDERLICH, J., CENTRAL SCI. INDUSTR. RES. ORG., MELBOURNE, AUSTRALIA
YAKEL,H.L.,OAK RIDGE NATIONAL LAB.,OAK RIDGE,TENN.
ZAKRAJSEK, E., COMPUTING CENTRE, LEPI POT 11, LJUBLJANA, JUGOSLAVIA
ZAKRAJS, SEE ZAKRAJSEK, E.
ZALKIN, A., LAWRENCE RADIATION LAB., P O BOX 808, LIVERMORE, CALIF.
ZANOVELLO RENATO, C.E.C.S., V. PAOLOTTI 9, PADOVA, ITALY
ZANO, SEE ZANOVELLO RENATO
ZELENKO,B., INST.R.BOSCOVICH BIJENICKA 54 , ZAGREB, JUGOSLAVIA
ZILAHI-SEBESS, L., M.T.A.KOZP.KEM.KUT.INT., BUDAPEST, HUNGARY
ZUCCA, T., SORIN, SALUGGIA, ITALY

CNRS BELLEVUE = LABORATOIRE DE CRISTALLOGRAPHIE APPLIQUEE DU CNRS, 1 PLACE
    ARISTIDE BRIAND, BELLEVUE, S ET O, FRANCE
DAW-STRUKTURFORSCHUNG, BERLIN, GDR = INSTITUT FUR STRUKTURFORSCHUNG DER DAW,
    BERLIN-ADLERSHOF, GERMAN DEMOCRATIC REPUBLIC
INST. OF SOLID STATE PHYS., CSAV, PRAGUE, CZECHOSLOVAKIA = USTAV FYSIKY
    PEVNYCH LATEK CSAV, PRAHA-STRESOVICE, CUKROVARNICKA 10, CZECHOSLOVAKIA
LAB.CALC.CNRS = LABORATOIRE DE CALCUL DU CNRS, IMPASSE D*AUBERVILLIERS, PARIS
    19, FRANCE
M.T.A.KOZP.KEM.KUT.INT., BUDAPEST, HUNGARY = CENTRAL RESEARCH INSTITUTE
    OF CHEMISTRY OF THE HUNGARIAN ACADEMY OF SCIENCES, BUDAPEST, HUNGARY
NUM.CENTR.OF MAT., CHARLES UNIV., PRAGUE, CZECHOSLOVAKIA = CENTRUM NUMERICKE
    MATEMATIKY KARLOVY UNIVERSITY, PRAHA-MALA STRANA, CZECHOSLOVAKIA

LIST OF PROGRAMS

LISTED ACCORDING TO MACHINE TYPE AND FUNCTION.

TO FIND A PROGRAM WHEN ONLY THE ACCESSION NUMBER IS KNOWN, CONSULT THE
ACCESSION NUMBER INDEX WHICH GIVES THE MACHINE AND FUNCTION UNDER WHICH THE
PROGRAM IS LISTED.

IN THE INTEREST OF LEGIBILITY, THE ACCESSION NUMBERS AND SERIAL LETTERS (A, B,
ETC.) ON THE ABSTRACT CARDS (COLS. 1 - 5) HAVE BEEN SUPPRESSED.

CRYSTALLOGRAPHIC PROGRAM SYSTEMS AND PROGRAMS BELONGING THERETO --

3068 ATSYS   SYSTEM   DREW   ATSYS/ ATLAS CRYSTALLOGRAPHIC SYSTEM
    IMPERIAL COLLEGE SYSTEM OF CRYSTALLOGRAPHIC PROGRAWORKING IN CON-
    JUNCTION WITH A SINGLE MAGNETIC TAPE WHICH CONTAINS ALSO THE DATA AND
    WORKING SPACE FOR 20 - 30 JOBS. ALL PROGRAMS WRITTEN IN EXCHLF. THEY ARE
        FIFI DATA REDUCTION, WEISSENBERG (DIAMAND, DREW).
        PLIP DATA RECUTION, PRECESSION (DOLLIMORE, DREW)
        LSCD LS FIT POWD. OR S.C. DATA FOR CELL DIM. (DIAMAND)
        POLO SCALES DATA ON TWO OR MORE AXES (DREW)
        PEPE FOURIER (HARDING, DREW)

```
        KIKI   PAPER TAPE TO MAG TAPE DATA STORAGE (DREW)
        BABA   BLOCK DIAG. SF LS, SIMILAR TO ROLLETT,S MERCURY PGM(DIAMAND,DREW)
        ELSI   BOND DISTANCES AND ANGLES (DIAMOND, DREW)
        DIDO   WEIGHTED LS PLANE THRU GROUP OF ATOMS (SPARKS, HUNT)
        DIAN   DIHEDRAL ANGLES, ETC., FOR MODEL BLDG. (DIAMAND)
     FOR DETAILS CONCERNING THE SYSTEM, WRITE TO MR. M. G. B. DREW,
     IMPERIAL COLLEGE, UNIVERSITY OF LONDON, LONDON S.W. 7, ENGLAND
```
413 X-RAY-63 GENERAL   JSTEW*HIGH,HOLDEN,CHAST,DICSN,BROWN,DAYH,MOROSIN,LING LWA
    ,JENSEN,TAKEDA,WAX,KEEFE,AMMON,BRAUN,ALDEN/EACH PROGRAM OF THIS SYSTEM IS
    ALSO LISTED BY ITS FUNCTION. MAJOR CODING IS IN FORTRAN II WITH NEEDED
    FAP SUBROUTINES. SYSTEM IS EXECUTABLE ALONE OR UNDER ANY OTHER MONITOR.
    WRITE-UP AVAIL AS 1401 PROGRAM, SYMBOLIC DECKS ALSO FROM 1401.
    ALL ON MAG. TAPE OF SPECIFIED DENS. 2 REELS OF TAPE 8K IBM 1401 REQ.
    FORTRAN IV-NPL VERSION IN PREPARATION.
419 X-RAY-63 CONTOUR   DAYHOFF,STEWART*/CONTUR/ CONTOUR MAP LINK   LWA
    CONTOUR PLOTS OF A THREE DIMENSIONAL FOURIER MAP ARE PRODUCED. UP TO
    17,000 POINTS PER LAYER CAN BE HANDLED. THE LOGIC IS DESCRIBED IN
    COMM. ACM 6, (1963), 620-622
423 X-RAY-63 DIR   DICKINSON/JSTEW ESORT/PROGRAM TO SORT E VALUES   LWA
    (QUASI-NORMALIZED STRUCTURE FACTS.) PRELIMINARY TO KARLE SYMBOLIC
    ADDITION PROCEDURE.
440 X-RAY-63 DP   CHASTAIN/JSTEW DATCO1/SPECTROMETER DATA SCALING AND   LWA
    INTENSITY CORRECTION(INCLUDING BURNHAM GEN.ABS.CORR.)
439 X-RAY-63 DP   CHASTAIN/JSTEW DATCOR/INTER-FILM DATA SCALE AND AVE   LWA
443 X-RAY-63 DP   CHASTAIN/JSTEW IFORMS/GEN REF NEEDED + FORMS FOR   LWA
    RECORDING INTENSITIES FROM FILM DATA AS WELL AS INPUT TO DATCOR LINK.
441 X-RAY-63 DP   HOLDEN/JSTEW   DATCO2/INTER-FILM DATA SCALING   LWA
421 X-RAY-63 DPDIR   HOLDEN,STEWART* DATFIX/CONVERSION OF F-RELATIVE   LWA
    INTO QUASI-NORMALIZED STRUCTURE FACTORS(E-FACTORS)(KARLE-HAUPTMAN)
    THESE MAY THEN BE USED FOR THE SYMBOLIC ADDITION PROCEDURE OR FOR
    COEFFICIENTS IN SHARPENED ORIGIN REMOVED PATTERSON (VECTOR-MAP) OR
    E-MAPS BY THE FOURR LINK. ESTIMATE OF OV T.F. AND SCALE GIVEN.
438 X-RAY-63 DP   JSTEW*CHAST,MOROSIN/DATRDN/THIS LINK OF THE X-RAY SYS LWA
    SERVES TO BUILD THE COMPOUND REFLECTION TAPE WITH ALL SYMMETRY AND CELL
    INFORMATION STORED ON IT. ITS FUNCTIONS ARE PRELIMINARY TO ALL OTHER
    MAJOR LINKS OF THE SYSTEM. IT CALCULATES 1/LP FOR POW,WEIS,PREC,SPECTROM,
    IT DOES BOND ABS CORR.,INTERPOLATES FORM FACTOR FROM LITT BY 4-POINT,
    SETS LEAST SQUARES WEIGHTS,AND GENERALLY SCANS PRELIMINARY DATA. MOST
    IMPORTANT FUNCTION IS BUILDING OF FOURIER AND LEAST SQUARES GEN SYM CODE
442 X-RAY-63 DP   TAKEDA,JSTEW*   INCOR/WEIS INTEN.CORR.1/LP,BURNHAM   LWA
    GENERAL ABSORPTION CORR.
418 X-RAY-63 FR   HIGH,JENSEN,HOLDEN,JSTEW*FOURR/3D FOURIER,ALL SPC GRP LWA
    ALL SETTINGS,ANY INTERVAL TO 2000THS IN 1ST SUM,400THS IN 2ND SUM 200THS
    IN 3RD SUM DIRECTION,14 POSS COEF.(FO,DELTAF,E,FC,ETC.)WILL GIVE UNDISTOR
    SECTIONS,BINARY OUTPUT TO PLOTTER AND PEAK SEARCH LINKS OF SYSTEM. WILL
    TAKE VERY LARGE CELLS WITHOUT MODIFICATION.
431 X-RAY-63 FRBLK   HOLDEN/STEWART FOUREF/BLOCK FOURIER ATOMIC COORD REF LWA
    MOVES ATOMS TO MAXIMUM IN ED.CALCS,FC AND BLOCK,CYCLES,SHIFTS OVERALL TF.
    ANY SPACE GROUP
433 X-RAY-63 ID   CHAST,HIGH,JSTEW*BONDLA/GENERAL LENGTH ANG.GEN H COOR LWA
434 X-RAY-63 ID   BUSING,MARTIN,LEVY/STEWART/ORFFE/X-RAY-SYSTEM COPY   LWA
427 X-RAY-63 LC   ALDEN,AMMON,CHAST,HIGH/JSTEW*/PARAM/ REFINE CELL
    CONSTANTS FROM THETA VALUES BY LEAST SQUARES
416 X-RAY-63 LSBLK   STEWART*DICSN   BLOKLS/BLOCK DIAGONAL LEAST SQUARES   LWA
     MIXED T.F. 200 ATOMS,850 PARAM,UNLIM REF,ALL SPACE GROUPS SOME PATCH REQ
415 X-RAY-63 LSDM   STEWART*LING   DIAGLS/DIAGONAL LEAST SQUARES MIXED T LWA
     F.,250 ATOMS,1400 PARAM, UNLIM REF,ALL SPACE GROUPS,SOME SPEC POS PATCH
417 X-RAY-63 LSFM   STEWART*/EXTENSIVE MOD OF BUSING,MARTIN,LEVY F.M.LEAS LWA
     SQUARES,45 ATOMS,190 PARAM,UNLIM REF,ALL SPACE GROUPS,FIXED AT.SOME SPEC
     POS.PATCH REQ.MIXED T.F.REQ X-RAY SYSTEM REFLECTION TAPE.
436 X-RAY-63 LSPL   CHASTAIN/STEWARTLSQPL/METHOD OF SHOMAKER ET.AL. FITS   LWA
    AN ARRAY OF POINTS TO A LINE OR PLANE,GIVES INTER PL ANGS, DISTS FROM PL.
432 X-RAY-63 PATSHL   HOLDEN/STEWART SHLPAT/GIVES SHELLS AS MERCATOR PROJ   LWA
437 X-RAY-63 SPEC   CHASTAIN/STEWARTDELSIG/LEAST SQUARES WEIGHT ESTAB-   LWA
    LISHMENT. EST EXTINCTION IN DATA PLOT FO/FC AS FN OF FO OR SNTH/LAMBDA
420 X-RAY-63 PK   DAYHOFF,STEWART*/PEKPIK/ PEAK PICKING LINK   LWA

THE PRECISE COORDINATES AND DENSITIES OF THE PEAKS IN A 3 DIMENSION-
AL FOURIER MAP ARE CALCULATED. UP TO 20,000 POINTS PER LAYER CAN BE
SEARCHED. SYMMETRICALLY UNIQUE PICKS ARE LISTED. THREE ORTHOGONAL PRO-
JECTION PLOTS OF ALL PEAKS IN ONE UNIT CELL ARE MADE ON THE 1401.
```
414 X-RAY-63 SF        STEWART*HIGH      FC/GENERAL STRUCTURE FACTORS ALL SPAC  LWA
    GROUPS ALL SPECIAL POSITIONS,MIXED T.F.,DISPERSION,POPULATION OF SITE,ETC
    2400 ATOMS/CELL OV.TF.2000 AT/CELL IND.ISO.TF.1000 AT/CELL IND.ANIFO.TF.
445 X-RAY-63 SPEC      CHASTAIN/JSTEW    WUUD/4K1401 SYS WRITE UP PRINTING      LWA
422 X-RAY-63 SPEC      STEWART           MODIFY/SEARCH AND MODIFY REFLECTION    LWA
    TAPE. GENERATE PSEUDO-DATA OF KNOWN ERROR FOR TEST PURPOSES.
430 X-RAY-63 SPEC      STEWART           STUJOB/GENERATE 2D STUDENT PROBLEMS    LWA
425 X-RAY-63 SPEC      TAKEDA,DONNAY/JSTEW/TRXLS/TRANSFORMATIONS OF CRYST       LWA
    SETTINGS AND SPACE GROUP SYMBOLS
444 X-RAY-63 SPEC      WAX,STEWART*      UDATE/8K1401 SYS MAINTENANCE           LWA
424 X-RAY-63 STAT      HIGH,STEWART*     FCLIST/LIST STRUCTURE FACTORS          LWA
426 X-RAY-63 STAT      KEEFE,STEWART*    RLIST/LIST R VALUES FOR ZONES ETC      LWA
435 X-RAY-63 STERO     BRAUN,HIGH/JSTEWPROJCT/ORTH OR PERSP PROJ OF MOL ON      LWA
    ARBITRARY PLANE
428 X-RAY-63 THETASET  FREER,AMMON,CHAST,HIGH/JSTEW*/GESET/ GENERATION OF       LWA
    SPECTROMETER SETTINGS FROM CELL DATA AND ORIENTATION INFORMATION
429 X-RAY-63 WEISDP    HOLDEN/JSTEW      FLMSET/MM COOR OF REF ON WEIS FLMS     LWA
```

PROGRAMS LISTED UNDER PROGRAMMING LANGUAGES (PSEUDO-MACHINES) --

```
8039 ALGOL60   ABS       DUISENBERG        UTR.D1/ABSORPTION COEFFICIENT         LPA
     ABS.CORR. FOR CONVEX POLYHEDRON OR ARBITRARILY SHAPED CRYSTAL
     AND CORR. FOR INHOMOGENEOUS X-RAY BEAM. X1 8K
5051 ALGOL 60  ABS LP    SCHULZ            ABS, LP  FOR EQUI-INCL.-WEIS          LWA
     PROGR. CONVERTS FILM-DENSITIES TO INTENSITIES. THESE INTENSITIES ARE
     MULTIPLIED WITH THE LP-FACTOR AND ABSORP. FACTOR FOR SPHERICAL SPECIMENS.
     TESTED ON TR4. STORAGE ABOUT 1500. INPUT (H, K, L, DENSI). OUTPUT
     (H, K, L, INTENSI)
5048 ALGOL 60  ABS LPPK SCHULZ             ABS, LP, PK-COR. EQUI-INCL.-WEIS      LWA
     PROGR. CONVERTS FILM DENSITIES TO INTENSITIES. THESE INTENSITIES ARE
     MULTIPLIED WITH THE LP-FACTOR AND ABSORPT.FACTOR FOR SPHERICAL SPECIMENS.
     PEAK-INTENSITIES ARE CONVERTED TO INTEGRAL INTENSITIES BY THE METHOD OF
     PHILLIPS (ACTA CRYS (1954), 7, 746-751). TESTED ON TR4. STORAGE ABOUT
     2000. INPUT (H, K, L, DENSI). OUTPUT (H, K, L, INTENSI)
 236 ALGOL     D         EICHHORN          RECW/RECIPR.TRANSL.FROM WEISSBG.DATA  LPM
     PROGRAM COMPUTES RECIPROCAL TRANSLATIONS FROM AXIAL DATA ON WEISSENBERG
     FILMS, AND OPTIMIZES FOR FILM SHRINKAGE AND SIMILAR ERRORS.
 248 ALGOL     DF        EICHHORN          DSBM/DIFF.SYNTHESIS(DS-ONE)           LPM
     SIMPLE 3-BY-3 PLUS 6-BY-6 MATRIX DIFFERENTIAL SYNTHESIS FOR P-BAR-ONE,
     FOR UNLIMITED NUMBER OF REFLEXIONS BUT NUMBER OF ATOMS AND SCATTERING
     TYPES LIMITED BY AVAILABLE MEMORY. GOODNESS-OF-FIT AND ACCURACIES ARE
     COMPUTED SIMULTANEOUSLY WITH REFINEMENT. PROGRAM IS SELF-CYCLING.
 249 ALGOL     DF        EICHHORN          DSSO/DIFF.SYNTHESIS(DS-TWO)           LPM
     PROGRAM AS DS-ONE, BUT COMPUTES TAYLOR EXPANSION OF POSITIONAL ERROR TO
     SECOND ORDER APPROXIMATION.
 250 ALGOL     DF        EICHHORN          DSEI/DIFF.SYNTHESIS(DS-THREE)         LPM
     PROGRAM AS DS-TWO, BUT ALSO COMPUTES DEBIJE-WALLER SHIFTS TO SECOND ORDER
     CORRECTIONS. A RELAXATION METHOD IS APPLIED TO SOLVE EQUATIONS.
 251 ALGOL     DF        EICHHORN          DSTT/DIFF.SYNTHESIS(DS-FOUR)          LPM
     HAS ALL FEATURES OF DS-ONE, BUT IS BASED ON COMPARISON OF OBSERVED AND
     CALCULATED THIRD DERIVATIVES OF RHO, THUS YIELDING 10-BY-10 MATRIX. THE
     PROGRAM THUS CATERS TO INDIVIDUAL ATOMIC SCALE FACTORS.
8034 ALGOL60   DIR       BEURSKENS         UTR.B6/SIGNDETERM. FOR PROJECTIONS    LWA
     USING NONCENTR. 3DIM DATA   FORMULA B3,0 KARLE AND HAUPTMAN 1959  X1
8035 ALGOL60   DIR       BEURSKENS         UTR.BIO/SYMBOLIC ADDITION METHOD      MPA
     SIGNDETERM. FOR CENTR.STRUCT.  KARLE AND KARLE 1963, BEURSKENS 1964   X1
8032 ALGOL60   DP        BEURSKENS         UTR.B2/CALC. OF E-VALUES        X1    LWA
6038 ALGOL60   DP        JOHANSSON G       LR/X-RAY DATA ON    LIQUIDS           LPA
     COMPILED AND RUN ON FACIT. CORE 2K, DRUM 8K
7557 ALGOL     DP        PANATT/BOMB/COJAZ/SOME GEOMETRIC PROGRAMS               LPA
     COMP. ON 6001 1K.  1) BOND ANGLES AND DISTANCES, 2) BEST PLANE WITH LS.,
```

```
              3) GEOMETRICAL DETERMINATION OF THIRD COORD. FROM TWO KNOWN, 4) DP
              VALUES FOR TWO CIRCLE NEUTRON DIFFRACTOMETER
  8059 ALGOL60  DP WEISS GEISE,HESPER      OX1/INTENSITIES TO SF WILSON PLOT         LWA
              USES LP CORRECTION, PHILLIPS CORR., AND ABSORPTION FOR CYLINDER
              OPTIONAL PROD INPUT FOR OX3, OX4, OX5, OX8, OX9 . TESTED ON X1 8K.
   247 ALGOL     E ID     EICHHORN         ESDV/ESD AND MPE OF POS.BONDS,ANGLES      LPM
              FROM THE ESD VALUES OF X,Y,Z THE MOST PROBABLE ERRORS OF BONDS AND ANGLES
              ARE COMPUTED TOGETHER WITH THE BONDS AND ANGLES.
   256 ALGOL     FR       EICHHORN         FRSY/TRICLINIC FOURIER SYNTHESIS          LPM
              FOURIER SYNTHESIS PROGRAM (3-DIMENSIONAL) FOR TABULATOR PRINT-OUT.
              INDEX LIMITATION IS FACTOR OF AVAILABLE STORAGE.
  5012 ALGOL    FR 2      HOPPE            2 DIM NO SYMMETRY SIMPLIFICATION          LWA
  8028 ALGOL60  FR23 BL   PETERSE          H182/GENL ALL SPACEGROUPS                 LWA
              IT.IND.REFL.AND SPACEGR.DEPENDING PROC.TO GENERATE DEP.REFL,CONTOUR IF
              WANTED FROM PAPERTAPE OT,UP TO 4000 REFL. AGLTR4 8K
  8529 ALG     FR 2 3     UEDA,I.          UFO-1/FR GRID N/120 TRI MONO ORTH         LPA
              FOR TRI MONO ORTH. GRID N/120. CHOICE OF SECTION FREE. TWO VALUES
              OF ABS(HKL) LESS THAN 30. PATT,FOURIER,D-F POSSIBLE. INPUT-M.T.
              OUTPUT-PRINTS. PROG.TEST REFER SFU-1.
  6036 ALGOL60  H         WENGELIN         GENTRAI/GENERATES  TRANSFORMS INDICES LWA
              TRANSFORMS TO NEW CELL. GENERATES BY MAX 2 TWIST, 3 REFLEXION-MATRICES.
              CHECKS AND SURPRESSES IDENTICAL REFLECTIONS WITHIN EACH GENERATION.
              COMPILED AND RUN ON 16K DATASAAB D21.
   240 ALGOL    H LC      EICHHORN         CELL/CELL CONSTANTS,LIMITING SPHERE       LPM
              PROGRAM COMPUTES COMPLETE SET OF CELL DATA FROM OBSERVED DATA ABOUT TWO
              AXES, THEN COMPUTES ALL NON-REDUNDANT INDEX COMBINATIONS SUBJECT TO
              SPACEGROUP CONSTRAINTS WITHIN GIVEN LIMITING SPHERE (WAVE LENGTH).
  5049 ALGOL 60 ID        SCHULZ           DIST.+ANGLES BY GIVEN KOORD.              LWA
              DISTANCES OR ANGLES ARE CALCULATED BY INPUT OF KOORD. TRIPLES (2 OR 3).
              TESTED ON TR4. STORAGE ABOUT 300.
  6041 ALGOL60  ID        NYBORG.DANIELS*  ND1/COMPLETE SET OF DISTANCES.            LWA
              GIER AGL FAST STORE 1024 WORDS. DRUM STORE 12800 WORDS. SPEC INSTRUCTIONS
              FOR DRUM OPERATIONS. ALL RELEVANT DISTANCES AND ANGLES BETWEEN ATOMS IN
              27 UNIT CELLS. STANDARD DEVIATIONS.
   409 ALGOL    ID E      GALLAHER,TAYLOR  ALGOL FUNCTION AND ERROR                  LWA
              ALGOL TRANSLATION OF OAK RIDGE FORTRAN FUNCTION AND ERROR BY BUSING,
              MARTIN, AND LEVY. I/O SUITABLE FOR BURROUGHS B5000 AND B5500
   241 ALGOL    LC        EICHHORN         ROTO/AXES FROM LAYERLINES                 LPM
              PROGRAM CALCULATES AXIAL TRANSLATIONS FROM ROTATION-OSCILLATION DATA,
              AND OPTIMIZES FOR FILM SHRINKAGE ERRORS AND NON-CYLINDRICAL CAMERA SHAPE.
   243 ALGOL    LF        EICHHORN         SCCS/SCATTERING CURVES                    LPM
              PROGRAM CONVERTS PUBLISHED SCATTERING DATA INTO UNIFORM SET OF 100
              VALUES AND DIFFERENCES, TO GIVE CARD SET OF 25 CARDS FOR INPUT INTO
              STRUCTURE FACTOR AND REFINEMENT PROGRAMS LISTED BELOW.
   238 ALGOL    LP        EICHHORN         WEIS/LORENTZ-POL.WEISSBG.DATA             LPM
              PROGRAM CALCULATES LP FACTOR ACCORDING TO 3-DIMENSIONAL EXPRESSION AND
              OBTAINS BOTH PATTERSON AND FOURIER COEFFICIENT FROM RAW INTENSITY.
   244 ALGOL    LP        EICHHORN         PREC/LORENTZ-POLAR.PRECESSION DATA        LPM
              PROGRAM USES THE MODIFIED WASER EXPRESSION TO OBTAIN LP FACTORS FOR
              PRECESSION INTENSITIES. THESE ARE CONVERTED INTO THE PATTERSON AND
              FOURIER COEFFICIENTS.
  7548 ALGOL    LS        BASSI            1/X.RAYS/NEUTRONS OBSREV.                 LPN
              LEAST SQUARES REFINEMENT,FM,ISOTROPIC AND ANISOTROPIC TF,ALL SPACE GROUPS
              LF LINEAR INTERPOLATION,E,R,SCL, CORRELATIONS MATRIX,INTERAT.BOND LENGTHS
  7549 ALGOL    LS        BASSI            2/X RAYS AND NEUTRONS MIXED OBSERV.       LPN
              LEAST SQUARES REFINEMENT,FM,ISOTROPIC AND ANISOTROPIC TF,ALL SPACE GROUPS
              LF LINEAR INTERPOLATION,E,R,SCL, CORRELATIONS MATRIX,INTERAT.BOND LENGTHS
   252 ALGOL    LS        EICHHORN         LSAN/LEAST SQUARES REF.(LS-ONE)           LPM
              LEAST SQUARES ANALOG OF DS-ONE.
   253 ALGOL    LS        EICHHORN         LSFM/LEAST SQUARES REF.(LS-TWO)           LPM
              HAS ALL OPTIONS OF DS-ONE AND LS-ONE, BUT COMPUTES 9-BY-9 MATRIX.
  6037 ALGOL60  LS        JOHANSSON G      FLR/REFINEMENT X-RAY DATA ON LIQUIDS      LPA
              COMPILED AND RUN ON FACIT. CORE 2K, DRUM 8K
  6039 ALGOL60  LS,AT     GROENBAEK        3X3,6X6 BLOCK DIAGONAL LS                 LPA
              GIER AGL. FAST STORE 1024 WORDS. DRUM STORE 12800 WORDS.SPEC
              INSTRUCTIONS FOR DRUM OPERATIONS. PROGRAM MODIFICATIONS FOR MOST
```

```
                SPACE GROUPS AND SPEC POS.
  410 ALGOL      LS AT FM GALLAHER,KAY      ALGOL LEAST SQUARES SF,FM,AT       LWA
      ALGOL TRANSLATION OF OAK RIDGE FORTRAN LEAST SQUARES BY BUSING, MARTIN,
      AND LEVY. I/O SUITABLE FOR BURROUGHS B5000 AND B5500
 8024 ALGOL60    LS LC P   VISSER JW        H932/TRICL.MONOCL. ORTHOR.         LPA
      AGLTR4,IT.RECIPR.LC,OBS.REFL.OT.LC,LIST OF OBS.REFL.AND ALL CALC.REFL.,IN
      CREASING THETA
 5052 ALGOL      LP PREC   BAERNIGHAUSEN    LP CORRECTION FOR PREC DATA FOR ANY LWA
      LAYER PARALLEL A*,B*. IT RECIPROCAL LC, LAYERINDEX L, PREC-ANGLE,
      HK AND INTENSITY. OT SF AND SQUARE OF SF.
 5072 ALGOL 60   P THTA    D VON STEINWEHR  ARRAU 170565                       LWA
 8025 ALGOL60    PDIAGRAM  VISSER JW        H627/POWDERDIAGRAM FROM LITERAT.DATA LPA
      ALGTR4,CALCUL.INTEGR.INTENS.,IT.RECIPR.LC,POS.TF,OT.DIAGRAM FOR ANY INSTR
 7551 ALGOL      PI        BASSI            H BETWEEN 0 AND THETA MAX.         LPN
 8030 ALGOL60    PI        PETERSE          H81/Q-DIFFERENCES  NOT-TRICL.SYSTEMS LWA
      AGLTR4 IT.Q-VALUES,OT.FREQUENCY OF DIFFERENCES
 8027 ALGOL60    PI 2,3    KUYPERS/PETERSE  H142,H127A/ZONEMETHOD GENL
      IT.Q-VALUES,2D DETERMINES INDEP.ZONES,3D COMBINES 2 ZONES TO RECPR.LATT.
      AGLTR4
  246 ALGOL      PROF      EICHHORN         LSPL/MARSH-WASER LEAST SQU.PLANE   LPM
      THE MARSH-WASER RECIPE IS APPLIED TO COMPUTE THE (OPTIMIZED) LEAST
      SQUARES PLANE THROUGH A SET OF POINTS.
 6040 ALGOL60    REFINEM.  DANIELSEN        D45/R MINIM BY TEST FOR EACH PARAM. LWA
      GIER AGL FAST STORE 1024 WORDS. DRUM STORE 12800 WORDS. SPEC INSTRUCTIONS
      FOR DRUM OPERATIONS. REFINEMENT BY SYSTEMATIC VARIATION OF THE PARAMETERS
      METHOD BY BHUIYA AND STANLEY, ACTA CRYST 16,981(1963). ISOTROPIC TF.
  239 ALGOL      S         EICHHORN         THSC/THRESHOLD SCANNER             LPM
      PROGRAM DETERMINES THRESHOLD INTENSITY VALUE FOR EACH OBSERVABLE BUT
      NOT OBSERVED REFLEXION, FROM LP EXPRESSION AND MINIMUM VISIBLE INTENSITY.
  245 ALGOL      S         EICHHORN         FILM/ACCURACY OF FILM READINGS    LPM
      PROGRAM COMPUTES MOST PROBABLE ERROR FOR CELL CONSTANTS FROM ESTIMATED
      STANDARD DEVIATIONS OF INDIVIDUAL READINGS.
  257 ALGOL      SCF       HOLM             SCFA/PNL(N) AND Z-CONTR.           LPM
      HIGHLY SPECIALIZED INTEGRATION PROGRAM TO CALCULATE SELFCONSISTENT FIELDS
      FOR HARTREE FIELDS OF LANTHANIDES.
  258 ALGOL      SCF       HOLM             SCFB/Y(R) AND Y*(NL,R)             LPM
      HIGHLY SPECIALIZED INTEGRATION PROGRAM TO CALCULATE SELFCONSISTENT FIELDS
      FOR HARTREE FIELDS OF LANTHANIDES.
  259 ALGOL      SCF       EICHHORN         ATOM/WAVEFIELD INTEGRATION         LPM
      CURVEFITTING INTEGRATION OF WAVEFIELDS TO OBTAIN ATOMIC SCATTERING
      FACTORS FOR DIFFRACTION USES.
 7552 ALGOL      SF        BASSI            X.RAYS OR NEUTRONS  STRUCT.FACT.   LPN
      POWDER SYSTEMATIC COMPUTATION OF SF FOR BRAGG ANGLES FROM 0 TO A GIVEN
      MAXIMUM,OR ONLY FOR A LIST OF GIVEN H,K,L ,ISOTROPIC OR ANISOTROPIC
  254 ALGOL      SF        EICHHORN         ISOF/IND.ISOTROPIC STRUCTURE F     LPM
      SIMPLE STRUCTURE FACTOR AND PERCENTAGE DISCREPANCY CALCULATION FOR
      ISOTROPIC TEMPERATURE FACTORS.
  255 ALGOL      SF        EICHHORN         ANIF/IND.ANISOTR.STR.FACTORS       LPM
      SIMPLE STRUCTURE FACTOR AND PERCENTAGE DISCREPANCY CALCULATION FOR
      ANISOTROPIC TEMPERATURE FACTORS.
 8528 ALG        SF        UEDA,I.          SFU-1/SF 6KINDS 150ATOMS ISO-TF    LPA
      FOR TRI MONO ORTH 6KINDS 150ATOMS ISO-TF. SCALE OF FOBS,R FACOTRS
      OF 10 REGIONS ARE CALCULATED. INPUT-CARDS. OUTPUT-PRINTS,M.T.(TO
      BE USED FOR FOURIER).PROG.IS TESTED BY OKITAC-5090-H(8K) WITH
      3M.T.,1DRUM,CARD-READER,PRINTER.
 8026 ALGOL60    SF 1,2,3  VISSER J W       H628/BHUIYA-STANLEYMETH.ACTA 16,987 LPA
      AGLTR4,SINGLE CRYST.,NO LIMIT ON REFL.PARAM.POS.SCL,OVERALL-B 12K IF 3D
 5014 ALGOL      SF 3      HUBER/ANZENHOF.  3 DIM ANY PLANE WANTED             NNN
 5050 ALGOL 60   SF 3      SCHULZ           3 DIM INDIV. ISOTR. TEMP. + MULT.  LWA
      STRUCTURE FACTORS FOR INDIVIDUAL ISOTROPIC TEMPERATURE FACTORS AND
      INDIVIDUAL MULTIPLIERS ARE CALCULATED AND THE AGREEMENT FACTOR IS
      ESTIMATED. TESTED ON TR4. STORAGE ABOUT 2500.
 8029 ALGOL60    SFLS      PETERSE,PALM     H562/BLOCK DM AT E GENL R          LWA
      AGLTR4,SF DATA ON MAGNETIC TAPE,ALL SPACEGROUPS,NO LIMIT ON REFL.ANALYTIC
      AL LF,VARIABLE WEIGHTINGSCHEME,AUTOMATIC REF.OF COORD.AND AT OF ALL SPEC.
      POSITIONS,PARAMETERS.SCL,ANISOTR. AND ISOTR.TF,POSITIONS
```

```
8023 ALGOL60   SF  P      VISSER J W          H952/BHUIYA-STANLEY FOR POWDER ACTA 16  LPA
     ALGTR4,IT.OBS.INTEGRATED INTENS.OF ALL INSTRUM.,PROGR.ALSO HANDLES COIN
     CIDING LINES,PARAM.POSITIONS,SCL,OVERALL-B
 237 ALGOL     SF  S      EICHHORN            TRER/TRIAL AND ERROR STR.F           LPM
     THIS PROGRAM WILL ROTATE AND TRANSLATE A RIGID GROUP SYSTEMATICALLY
     AND COMPUTE THE STRUCTURE FACTOR SET FOR EACH MODEL.  THE TRIAL AND
     ERROR PROCESS IS SELF-OPTIMIZING AND WILL CONVERGE UPON A SET R-FACTOR.
8040 ALGOL60   SPC        DUISENBERG          UTR.D2/CRYSTAL POSITION AND SETTING  LPA
     INPUT  OBSERVED CHI AND PHI FOR 3 REFLEXIONS. CALC  EXACT POSITION OF THE
     CRYSTAL AND SETTING ANGLES FOR GENERAL-ELECTRIC AND NONIUS DIFFR. X1
7550 ALGOL     SPEC       BASSI               N POWDER PATTERNS CORRECTIONS        LPN
8031 ALGOL60   STAT       BEURSKENS           UTR.B1/WILSON PLOT, LS FIT    X1     LWA
8033 ALGOL60   STAT       BEURSKENS           UTR.B3/DISTRIBUTION FUNCTION OF E-VAL
     DETERM. OF A CENTER OF SYMM. RAMACHANDRAN AND SRINIVASAN 1959  X1
8036 ALGOL60   STAT       DE VRIES A          UTR.V5/MAKES INPUT FOR UTR.V7 AND V9 LWA
     2DIM TABLE OF FO-FC OR FC/FO AGAINST I AND THETA FROM FO AND FC. X1 8K
8037 ALGOL60   STAT       DE VRIES A          UTR.V7/EXTINCTION CONSTANT           LWA
     CALC. EXTINCTION, SCF AND TF CORRECTION FROM FC/FO   SEE UTR.V5
8038 ALGOL60   STAT       DE VRIES A          UTR.V9/LS WEIGHTING SCHEME           LWA
     ANALYSIS OF FO-FC TABLE  SEE UTR.V5  IN TERMS OF THE PRODUCT OF A
     FUNCTION OF I AND A FUNCTION OF THETA  SEE ACTA CRYST 18,1077
8060 ALGOL60   STEREO     GEISE, RUTTEN       OX12/CALC. MOLECULAR GEOMETRY        LWA
     CALC ANGLES NON INTERSECTING LINES, ANGLES BETWEEN PLANES, AND LINE AND
     PLANE.  CALC LS PLANES AND DIST TO PLANES.  TESTED ON X1 8K.
8061 ALGOL60   STEREO     GEISE, RUTTEN       OX20/CALC.HYDR.POS. (TEST ON X1 8K)  LWA
8041 ALGOL60   TF         VAN EIJCK           UTR.E1/VIBRATIONAL ELLIPSOIDS        LWA
     CALCULATES VIBRATIONAL ELLIPSOIDS FROM GIVEN U-VALUES
 242 ALGOL     TF  S      EICHHORN            ZERO/ZERO-APPR.ANISOTR.TEMP.FACTORS  LPM
     PROGRAM CONSTRUCTS ZERO-APPROXIMATE SET OF ANISOTROPIC DEBIJE-WALLER
     PARAMETERS FROM CELL CONSTANTS AND OPTIMIZED ISOTROPIC TEMP.FACTORS.
 502 BALGOL    THETA      RUDEE , M.L.        CALC PK CENTROID BY PIKE WILSON METH.LWA
     CALCULATES CENTROID OF A PEAK BY THE METHOD OF PIKE AND WILSON, BRIT. J.
     APPL. PHYS., V.10, P.57, (1959).
 466 FTN 1410  ABS        SCHAPIRO            ABS2/POLYHEDRAL XTALS ON XRD-5       LWA
     CARD INPUT FROM DP2, CARD OUTPUT FOR RECLP
 450 FORTRAN2  ABS 3      BERNARD,LANGHMR     EXACT ABS CORR. PARALLELOGRAM CRYST. LPA
     FOR EQUI-INCL WEIS.  FOR MONOCL, ORTHO, TETR, CUB, NEEDLE CRYSTALS HAVING
     PARALLELOGRAM CROSS SECTIONS, ROTATING ABOUT NEEDLE AXIS. 32K REQUIRED.
     TESTED AT 25 SEC/REFLECTION ON 60K 1620. USED FOR MU=378.
 362 FORTRAN2  ABS 3      BUSING*LEVY         ORABS/ABSORPTION CORR,ARBITRARY SHAPE LWA
     USES GAUSS INTEGRATION TO CALCULATE ABSORPTION FACTOR FOR CRYSTAL DEFINED
     BY PLANE FACES. GENERAL THREE-DIMENSIONAL REFLECTION DESCRIBED BY ANGLES
     READ BY SUBROUTINE WHICH MAY BE WRITTEN BY USER. SUBROUTINES INCLUDED FOR
     ONE ZONE AND FOR GE ORIENTER. USES TAPE STORAGE IF NEEDED. WRITTEN IN
     FORTRAN 2. TESTED ON 704, 7090, AND 1604A.
 225 FORTRAN2  ABS EXT    HAMILTON            ESAF-GONO9/ABS,EXT,POLYHEDRAL XTL    LW1
     CALCULATES ABSORPTION AND SECONDARY EXTINCTION CORRECTIONS FOR POLYHEDRAL
     CRYSTALS IN GONIOSTAT GEOMETRY.  MAIN PROGRAM CAN BE RECOMPILED FOR OTHER
     GEOMETRIES.  SECONDARY EXTINCTION CORRECTION IS VALID EVEN FOR SEVERE
     EXTINCTION, SINCE THE COUPLED DIFFERENTIAL EQUATIONS OF THE DARWIN
     TYPE ARE SOLVED IN A GENERAL WAY.  TESTED  ON IBM 7094 . RUNNING
     TIME TYPICALLY TWO SECONDS PER REFLECTION. SEE ACTA CRYST 16,609.
 449 FT2       CONTOUR    DAYHOFF             CONTOUR MAP SUBROUTINES  (7094)      LWA
     USER WRITTEN MAIN PROGRAM STORES MATRIX AND CALLS CONTOUR MAP SUB-
     ROUTINES ONCE FOR EACH CONTOUR LEVEL. SUBROUTINES PRODUCE PLOT POINTS
     FOR EACH CONTOUR CURVE. USER WRITTEN PLOT ROUTINE IS CALLED EACH TIME
     POINTS FOR ONE CURVE ARE READY. SEE COMM. ACM 6 (1963), 620-622
3044 FORTRAN2  DEBYE      HALL                FOBS-FCAL FIT FOR TEMP FAC,EXT PARAM LWA
     FITS FOBS AND FCAL TO GIVE BEST SCALE,TEMP. FACTOR AND EXTINCTION
     PARAMETER. CORRECTIONS ARE APPLIED TO DATA AND R IS CALCULATED
 472 FTN 1410  DEBYE      SCHAPIRO            THERMAL/PR.AXES ,RMS VIB. AMPL.      LWA
7535 FORTRAN4  DP         DOME,VACIAGO        PRFR/PREPAR. OF FOURIER AMPLITUDES   LNA
     THE PROGRAM PREPARES FOURIER (OR DIFFERENCE FOURIER) AMPLITUDES FROM THE
     OUTPUT OF SF PROGRAM (FAST). CAN ALSO BE USED TO PREPARE PATTERSON
     AMPLITUDES. AVAILABLE AT PRESENT IN A TEMPORARY VERSION FOR SPACEGROUPS
     1,2,10,12,14,15,19,29,47,53,61,62, BUT DEVISED FOR ALL TRICLINIC,
```

```
                MONOCLINIC AND ORTHORHOMBIC SPACEGROUPS. TESTED ON 16K 7040.
 465 FTN 1410 DP          SCHAPIRO          DP1,DP2/CALC.OF IOBS AND STD.DEV.IOBS LWA
                CARD INPUT OF XRD-5 DATA,CARD OUTPUT FOR PROGRAMS ABS2 OR RECLP
 476 FORTRAN  DP          TROTTER           DATAPR/GONIOSTAT SETTINGS, DATA PREP LWA
                PROGRAM GENERATES INDICES HKL AND COMPUTES SETTINGS FOR G.E. GONIOSTAT.
                APPLIES LP FACTORS FOR EQUI-INCLINATION WEISSENBERG FILMS, INTERPOLATES
                ON UP TO 8 SCATTERING CURVES. IBM 7040  32K
 120 FORTRAN2 DP      WRIGHT,CHAND,SHO*/DATAPR/SCL+TF,FCOEF,UNTRY F,PATCOEF LPA
                DOES WILSON PLOT (BY LEAST SQUARES) TO GIVE OVERALL ISOTROPIC TF AND
                PRELIM SCL FACTOR.  GIVES PATTERSON COEFS MODIFIED BY ANY DESIRED SHAPE
                FUNCTION AND WITH ORIGIN PEAK REMOVED.  THEN RECALCULATES SCL FACTOR BY
                CONSIDERATION OF PAT ORIGIN.  PUNCHES F (ABS SCL) FOR BUSING (PGM 360)
                AND PREWETT (PGM 364) LS PGMS. ALSO PUNCHES UNITARY SF CARDS. COMPILED
                ON 32K 7090, SHOULD RUN ON 8K MACHINE.
7023 FORTRAN  EL          TOURNARIE         304C/ELECTRON DYNAMICAL DIFFRACTION  MPA
 478 FORTRAN  FR          TROTTER           FOURIE/3-D FOURIER PROGRAM           LWA
                ANY SPACE GROUP. 1/120 THS OR MULTIPLES OF CELL EDGES. 16 X 31 X 31
                POINTS IN ONE PASS. IBM 7040 32K
 344 FORTRAN  FR2         KAY               IT A,B(CAL),FOBS.OT FR/DIF.          NPA
                CALCULATES EL. DENSITY OR DIFFERENCE MAP USING SIN. TABLE
                CHECKED ON 8000 WORD 704
 504 FORTRN63 FR 2, 3 HEATON,MUELLER  B114/GENL,BL,120 GRID (MIFRI)              LWA
                FORTRAN MODIFICATION OF MIFRI(PGM118) 64K REQUIRED. GENERALLY SIMILAR
                BUT MANY INPUT INSTRUCTIONS PERFORMED AUTOMATICALLY BY SUPPLYING
                INTL TABLES SPACE GROUP NUMBER(1-74). INPUT ALSO IN FORM OF ACTUAL
                DIFFRACTION COORDINATES. STRAIGHT OR DIFFERENCE FOURIER WITH VARIABLE
                FORMAT OUTPUT SHEETS FOR CONTOURING. 32K VERSION AVAILABLE AS B149.
7534 FORTRAN4 FR 3 BL  DOME,VACIAGO    FOUR/GENERAL 3-D FOURIER, GRID N/100 LNA
                BEEVERS-LIPSON PRINCIPLE. SIN, COS VALUES BY TABLE LOOK-UP. INPUT TAPE
                FROM PRFR. THE SORTING ORDER OF DATA ON THE INPUT TAPE DETERMINES ON
                WHICH AXIS THE SECTIONS ARE CUT. THE OUTPUT SHEETS PERMIT DIRECT
                CONTOURING. TESTED ON 16K 7040.
7030 FORTRAN  FT          BUJOSA                                                LPA
                THIS PROGRAM COMPUTES THE DIFFERENCE FOURIER TRANSFORM (DFT) OF MOLECULES
                IT HAS BEEN PROGRAMED FOR TRICLINIC CRYSTALS AND STRUCTURE FACTORS MAY BE
                EVALUATED AS WELL. THIS PROGRAM HAS BEEN WRITTEN FOR IBM 1620 DP SYSTEM
                PROGRAM STORAGE' 12296. WORK STORAGE' 2980.
 395 FORTRAN  FT, DFT ASHWORTH,CANUT   FT AND DIFFUSE SCATTERING DFT, TDS1/K LPA
                GENERATES FRACTIONAL HKL AND CALCULATES MOLECULAR FT, DIFFERENCE FT AND
                THE STRUCTURE DEPENDENT PART OF THE THERMAL WAVES DIFFUSE SCATTERING.
                FORTRAN II ON THE IBM 1620, 40K WITH A 1311 DISK.  FORTRAN IV ON THE IBM
                7040, 32K WITH 8 TAPES.
7025 FORTRAN  FT2 ID      TOURNARIE         307C/FOURIER TRANSFORM WHITH PICTURE MWA
 411 FORT2FAP GENERAL     STEWART*HIGH      X-RAY-63/SYSTEM FOR IBM 709-7094-(DC) LWA
                SEE ALSO UNDER X-RAY-63 ENTRY IN LIST FOR MORE DETAIL
7543 FORTRAN  HYDR GEN COL,ESP,GIG*IAN      HYDROGEN COORDINATES GENERATION     LNA
 506 FORTRAN  ID          BAUR              BOND LENGTHS AND ANGLES             LWA
                GIVEN THE ATOMIC COORDS AND EQUIVALENT POSITIONS IT SEARCHES FOR ALL THE
                INTERATOMIC BOND LENGHTS AND ANGLES WITHIN SPECIFIED LIMITS. 7090
7541 FORTRAN  ID          COL,ESP,GIG*IAN   ID AND ANGLES IN TRICLINIC CELL     LNA
                ALSO INTERMOLECULAR DISTANCES LESS THAN A FIXED DISTANCE,1620 20K
 396 FORTRAN4 ID,E        HARRIS,D.R.       DAESD/7044                          LPA
                INTERATOMIC DISTANCES AND ANGLES WITH THEIR EST STANDARD DEV. DISTANCES
                BETWEEN ASYMMETRIC UNIT AND ALL EQUIVALENT POSITIONS IN UNIT CELL AND 26
                NEAREST NEIGHBOR CELLS ARE GIVEN. OPTIONAL OUTPUT OF CART COORD (ANG).
3046 FORTRAN2 ID          OCONNOR           DISTS,ANGLES BETWEEN SPECIFIED POSNS LWA
                CALCULATES DISTS AND ANGLES BETWEEN SPECIFIED ATOMS ONLY, GIVEN THE
                FRACTIONAL COORDINATES
3045 FORTRAN2 ID          RAE               BOND DISTANCE SEARCH                LWA
                ALL INTERATOMIC DISTANCES LESS THAN A GIVEN MAXIMUM ARE CALCULATED
                FROM THE COORDINATES IN THE ASYMM. UNIT AND EXPRESSIONS FOR THE
                EQUIVT. POSNS.
 471 FTN 1410 ID          SCHAPIRO          BONDS/DIST. AND ANGLES FOR SEL. ATOMS LWA
 470 FTN 1410 ID          SCHAPIRO          VSET/ INTRA,INTER-MOLEC. VECT.AND DST LWA
 532 FORTRAN2 ID          SHOEMAKER, D.     DISTAN/DISTANCES,ANGLES,DIHEDRAL ANG. LPA
                CALCULATES INTERATOMIC DISTANCES, BOND ANGLES, AND DIHEDRAL ANGLES.
```

GIVES ALL PARAMETERS NEEDED FOR CRYSTAL MODEL CONSTRUCTION. ALL SPACE
GROUPS AVAILABLE, SPECIFIED BY SYMMETRY CARDS, ONE FOR EACH GENERATOR.
ALL NECESSARY INSTRUCTIONS ON COMMENT CARDS (NO SEPARATE WRITE-UP).
TESTED ON IBM 7090.

479 FORTRAN ID   TROTTER   CRRES/LENGTHS, ANGLES, MEAN PLANES ETC   LWA
COMPUTES BOND LENGTHS, VALENCY ANGLES, MEAN PLANES, INTERMOLECULAR
DISTANCES. IBM 7040 32K

7020 FORTRAN LC   BIBIAN,TOURNARIE/341A/OPTIMAL SPACING   MPA

386 FORTRAN2 LC P PI   HECKEL   POWD SOLN-CUBIC-MULTIPHASE PATTERN   LPA
MULTIPHASE POWD PATT DATA FITTED TO CUBIC HKL SEQ (BCC,FCC,SC,DC).
PGM HANDLES EXTRA LINES, ABSENT LINES, IDENTIFIES LATTICE(S), CALC LC,
EXTRAPOLATES LC TO COS SQ THETA EQUALS ZERO. PRINTS D SPACING, LC,
LATTICE TYPE(S), EXTR LC

226 FORTRAN LC   MUELLER,HEATON   B106/CUBIC THROUGH TRICLINIC   LWA
PROGRAM PERMITS THE DETERMINATION OF LATTICE CONSTANTS INCLUDING EDGES
AND ANGLES FOR ALL CRYSTAL SYSTEMS USING LEAST SQUARES. PROVISION MADE
FOR USING AS MANY AS THREE SEPARATE CORRECTION TERMS FOR ECCENTRICITY,
ABSORPTION, ETC.-HOWEVER, ONE, TWO, THREE OR NONE AT ALL MAY BE USED.
NO REITERATION IS NECESSARY.

463 FTN 1410 LC   SCHAPIRO   LSLAT/LS FIT OF WTED TWO THETA VALUES   LWA
COV.MAT.FOR REC. EL AND STD.DEV. CELL VOL

501 FORTRAN2 LC   WILLIAMS,D.E.   LCR2/LATTICE CONSTANT REFINEMENT   LWA
LEAST SQUARES TREATMENT OF FILM OR COUNTER DATA. ANY CRYSTAL SYMMETRY.
ECCENTRICITY, ABSORPTION, AND CALIBRATION ERRORS ARE CONSIDERED.

507 FORTRAN LP   BAUR   PREC ANY LEVEL   LWA
ANY SYMMETRY, ANY ORIENTATION OF CRYSTAL WITH RESPECT TO CAMERA. 7090

7536 FORTRAN LP   GIGLIO   GENERAL,FIVE KINDS OF ATOMS,PHILLIPS   LNA
INDICES TRANSFORMATION,ATOMIC SCATTERING FACTORS,1620 20K,NO S FEATURES

467 FTN 1410 LP   SCHAPIRO   RECLP/ LP FOR XRD-5   LWA
CARD INPUT FROM DP2 OR ABS2,CARD OUTPUT FOR   STAT1-4

3043 FORTRAN2 LPPREC   HALL   LP CORRECTIONS FOR PRECESSION METHOD LWA
GENERATES HKL AND LP FACTORS OR GIVES F AND FSQUARED FOR INPUT OF
HKL AND INTENSITIES

313 FORTRAN2 LS3 P FM HAMILTON   POWLS/GENERAL LS,OVERLAPPING P DATA   LWA
REFINES ARBITRARY FUNCTIONS OF SF FOR POWDER DATA, E. G., SUM OF
INTENSITIES OF TWO OR MORE REFLECTIONS. NON-DIAGONAL WEIGHT MATRIX
CAN BE USED. USER MUST SUPPLY SUBROUTINE FOR DEFINITION OF OBSERVATIONS
IN TERMS OF PARAMETERS. AVAILABLE DIMENSIONED FOR 50 OBSERVATIONS AND 20
REFINED PARAMETERS. ALSO AVAILABLE FOR 200 OBSERVATIONS AND 72 PARAMETER
ALTHOUGH THE LATTER VERSION USES DIAGONAL WEIGHTS ONLY. TESTED ON 7094.

528 FORTRAN LS AT   CARPENTER G   BXLS/GENL SF AND LS, BLOCK DIAG.   LWA
FLEXIBLE BUT SMALL SF AND LS. ADJUSTS COORDS., ISO. OR ANISO. TEMP.
PARAMETERS, ATOM SCALE FACTORS, AND OVERALL SCALE FACTOR. 10 X 10 BLOCK
DIAG. MATRIX. MAX. 24 INDEP. ATOMS, 36 EQUIV. POSITIONS.
COMPILED AND USED ON 5K 7070.

477 FORTRAN LS AT DM TROTTER   LS SQR/SFLS PROGRAM   LWA
STRUCTURE FACTOR LEAST SQUARES. UP TO ORTHORHOMBIC (EXTENDABLE TO HIGHER
SYMMETRIES). UP TO 60 ATOMS, 30 ANISOTROPIC. REFINES POSITIONAL AND
THERMAL PARAMETERS, AND SCALE FACTOR. STANDARD DEVIATIONS, NEW BOND
LENGTHS COMPUTED. IBM 704032K

390 FORTRAN LS AT FM LEVY,ELLISON* BMFLS/FTN LS,BIG MATRIX.SF,FM,AT,DSPN LPA
MODIFICATION OF XFLS, ACCESSION NO. 389. PROGRAM IS SEGMENTED TO ALLOW
FOR LARGER MATRIX. AS COMPILED ON 1604A (32K), MATRIX OF ORDER 243 CAN BE
USED. OTHER PROVISIONS SAME AS XFLS.

389 FORTRAN LS AT FM LEVY,ELLISON*   XFLS/FTN LS. SF, FM, AT, DISPERSN   LPA
MODIFICATION OF ORFLS, ACCESSION NO. 360. PROVISION FOR ANOMALOUS
SCATTERING, UNLIMITED NO. OF OBSERVATIONS, INFORMATION TRANSFER TO XFOUR
(AC 391), ORFFE(AC 363), ANALYZE(AC 392), EDIT(AC 393). CONDENSED
OUTPUT AND SEARCH OF CORRELATION MATRIX AVAILABLE. TESTED ON 1604A.
ACCOMODATES ORDER 187 MATRIX WITH 32K MEMORY.

7530 FORTRAN4 LS DM A   ALBA,DOME*VAC   MNQA/ANY SPACEGROUP, AT   LNA
THE SAME AS MNQC FOR ACENTRIC STRUCTURES. TESTED ON 16K 7040. PARAMETERS
OF ABOUT 70 ANISOTROPIC, 200 ISOTROPIC ATOMS CAN BE REFINED (WITH 16K).

7529 FORTRAN4 LS DM C   ALBA,DOME*VAC   MNQC/ANY SPACEGROUP, AT   LNA
BLOCK-DIAGONAL LEAST SQUARES FOR CENTRIC STRUCTURES. ISOTROPIC OR
ANISOTROPIC TEMPERATURE FACTORS CAN BE TREATED. BLOCKS 4X4 OR 9X9 PLUS

ONE BLOCK 2X2 FOR SCALE AND OVERALL B. EVERYTHING IS STORED IN HIGH-SPEED
MEMORY. TESTED ON 16K 7040. PARAMETERS OF ABOUT 85 ANISOTROPIC, 250
ISOTROPIC ATOMS CAN BE REFINED (WITH 16K).

7528 FORTRAN4 LS FM   ALBA,DOME*VAC   MNQF/ANY SPACEGROUP, AT   LNA
FULL MATRIX LEAST SQUARES. ISOTROPIC OR ANISOTROPIC TEMPERATURE FACTORS
CAN BE TREATED. EVERYTHING IS STORED IN HIGH-SPEED MEMORY. TESTED ON
16K 7040. REFINES UP TO ABOUT 110 PARAMETERS (WITH 16K).

7542 FORTRAN LS PLANE DAMIANI,GIGLIO* SCHOMAKER ET AL. METHOD,1620 20K LNA

469 FTN 1410 LSSF   SCHAPIRO   LSSF/AT,ICA,DISPERSN,FM OR BLOCK DM   LWA
VARIABLE SIZE DIAG.BLOCKS. CAN INCLUDE OVERAL SCALE IN EA BLOCK,CHANGE SF
BETWEEN BLOCKS. SYMMETRY INTERNAL FOR ANY POSN IN ANY SPACE GROUP
CALC.PAR.COV.MAT. AND STD.DEV. SUMS WTED ID AND ANGLES.   SF SUB-PROGRAM

7022 FORTRAN N   TOURNARIE   303B/POWDER DIAGRAM EXPLOITATION   MWA

345 FORTRAN N,DP,2 KAY   NPD AND SPECTR.IT FOR 100PART TABLE NPA
CHECKED ON 8000 WORD 704

385 FORTRAN2 N SF MAG KAY,CROMER   MAGNETIC STRUCTURE FACTOR   LWA
CALCULATES MAGNETIC STRUCTURE FACTORS FROM SPIN VECTORS AND ATOMIC
POSITIONS FOR EACH ATOM.

5061 FORTRAN2 PATSUP   ONKEN   QCOMPA/MOD OF SF FOR SHIFT PAT OR SUM LWA
QCOMPA MODIFIES SF. THE OUTPUT CAN BE PROCESSED BY ERFR2.THE COEFFICIENTS
ARE USED TO COMPUTE SHIFTED WEIGHTED PAT., SHARPEND PAT., SUM AND RELATED
FUNCTIONS USING UP TO 100 VECTORS AND SPACE GROUP SYMMETRY.

452 FORTRAN PATSUP   SIMPSON,LIPS*   HASUP   LWZ
USES ERFR2 BINARY OUTPUT AS INPUT. STORES ENTIRE PATTERSON MAP. COMPUTES
MINIMUM FUNCTION FOR UP TO 20 VECTORS.

482 FORTRAN PI   GOEBEL,WILSON   INDEX CUBIC POWDER PATTERNS   LWA
INDEXES POWDER PATTERNS ON CUBIC BASIS, USES MODIFIED LEAST SQUARES TO
CALCULATE MILLER INDICES, LATTICE CONSTANTS, STANDARD DEVIATIONS, ERROR
TERM AND FORMULA UNITS PER UNIT CELL. TESTED ON 7090. USED 5K STORAGE
LOCATIONS.

483 FORTRAN PI   GOEBEL,WILSON   INDEX HEX AND TETRAG POWD PATTERNS LWA
INDEXES POWDER PATTERNS ON HEXAGONAL AND TETRAGONAL BASIS. USES MODIFIED
LEAST SQUARES TO CALCULATE MILLER INDICES, LATTICE CONSTANTS, STANDARD
DEVIATIONS, ERROR TERM AND FORMULA UNITS PER UNIT CELL. TESTED ON 7090.
USES 5K STORAGE LOCATIONS.

484 FORTRAN PI   GOEBEL,WILSON   INDEX ORTHORHOMBIC POWDER PATTERNS LWA
INDEXES POWDER PATTERNS ON ORTHORHOMBIC BASIS. USES MODIFIED LEAST SQUARE
METHOD TO CALCULATE MILLER INDICES, LATTICE CONSTANTS, STANDARD DEVIATION
ERROR TERM AND FORMULA UNITS PER UNIT CELL. TESTED ON 7090. USES 5K
STORAGE LOCATIONS.

481 FORTRAN PI   GOEBEL,WILSON   INDEX ORTHO HEX TETRAG CUBE POWD PAT NNA
INDEXES POWDER PATTERNS ON CUBIC, HEXAGONAL, TETRAGONAL, AND ORTHORHOMBIC
BASIS, USES MODIFIED LEAST SQUARES TO CALCULATE MILLER INDICES, LATTICE
CONSTANTS, STANDARD DEVIATIONS, ERROR TERM AND FORMULA UNITS PER UNIT
CELL. TESTED ON 7090. USES 8K STORAGE LOCATIONS.

7032 FORTRAN4 PI   JAMARD   CTHO/CUB.TETRAG.HEX.ORTHO. POWD.INDX. LWS
CTHO MEANS CUBIC, TETRAGONAL, HEXAGONAL, AND ORTHORHOMBIC, IN WHICH
SYSTEMS THE PROGRAM INDEXES POWDER DIAGRAM, GIVEN EXPERIMENTAL U/D**2 AND
MAXIMUM EXPERIMENTAL ERROR IN SAME. PROGRAM CAN REJECT SOME IMPURITY
LINES.   TESTED ON 1107 UNIVAC

535 FORTRAN2 PI LS LC SHOEMAKER,C.B. LSCELP/INDEXES P,REFINES LC 7094 LPA
FOR ALL CRYSTAL SYSTEMS,INPUT 1/D OR THETA OBS AND
1/D CALC IN ORDER OF INCREASING 1/D

3047 FORTRAN2 PLC   LAWN   LS FIT FOR CUBIC CELL PARAMETER   LWA
CALCULATES CELL DIMENSIONS FROM POWDER LINE POSNS.USING WEIGHTED
LEAST SQUARES MODIFIED FOR NELSON-RILEY EXTRAPOLATION

7024 FORTRAN PROF   TOURNARIE   306H/SIZE DISTRIBUTION WHITHOUT T.F. MPA

5074 FORTRAN PROJECTN BIEDL,SHOEMAKER DISTAN-PRO
DISTAN-PRJCTN COMPUTES THE ORTHOGRAPHIC PROJECTION ON ANY PLANE (HKL)
OF AN ATOMIC STRUCTURE. IN ADDITION INTERATOMIC DISTANCES, BOND ANGLES,
AND DIHEDRAL ANGLES ARE COMPUTED. (THE DISTANCE-ANGLE PROGRAM *DISTAN*
WAS WRITTEN BY D. P. SHOEMAKER)

5060FORTRAN   S   GASSMANN   3DIMCONVOLUTION FOR BIG STRUCTURES   LWA
CONVOLUTION ACCORDING TO SAYRES EQUATION (AC5,52,P.60). DEVISED MAINLY
FOR PHASEDETERMINATION IN PROTEINSTRUCTURES (AC15,62,P.13) WITH MORE
THAN 1000 ATOMS AND UP TO 25000 UNIQUE REFLECTIONS (IBM 7090). ALL REFLEC

```
           TIONS ARE IN CORE STORAGE FOR MAXIMUM SPEED. PROGRAM WORKS IN ALL SPACE
           GROUPS. CALCULATIONS ON MYOGLOBIN HAVE BEEN PERFORMED.
 7028 FORTRAN  S         TOURNARIE          3120/X-RAY DYNAMIC IN BENDED CRYSTAL  MPA
 7533 FORTRAN4 SCL TF    DOME,VACIAGO       WILS/WILSON PLOT                      LNA
           THE PROGRAM PREPARES AND PRINTS OUT A WILSON PLOT FOR THE DETERMINATION
           OF THE SCALE FACTOR AND THE OVERALL B.
 7537 FORTRAN  SCL TF    GIGLIO             OBS INTENSITIES ON ABS SCALE,WILSON   LNA
 7532 FORTRAN4 SF AT     ALBA,DOME*VAC      FAST/ANY SPACEGROUP, UP TO 200 ATOMS  LNA
           GENERAL STRUCTURE FACTORS. ISOTROPIC OR ANISOTROPIC TEMPERATURE FACTORS
           CAN BE TREATED. CONTRIBUTIONS TO SF FROM NEW ATOMS CAN BE ADDED (OR
           SUBTRACTED) TO THOSE OF THE OTHER ATOMS. TESTED ON 16K 7040. THE PROGRAM
           ALSO CALCULATES SCALE FACTOR AND R.
  398 FORTRAN4 SFLS      HARRIS,D.R.        BDLSQ/AT,DISPERSN,BDM,ICA,7044        LPA
           SUBSTANTIAL REVISION OF SPARKS-TRUEBLOOD-OKAYA BLOCK DIAGONAL LS FOR
           IBSYS WITH MAP PACKING ROUTINE. MAX 18X18 MATRIX FOR TWO ATOMS, UP TO 6
           COORD AND UP TO 12 TF PARAMETERS PER MATRIX. 18X18 MATRIX FOR SCL AND
           FIRST 17 ISO TF. ALL SPACE GROUPS, 100 ATOMS, 2000 PLANES IN CORE.
 7021 FORTRAN  2SPC N    BIBIAN,TOURNARIE/323B/3 AXE SPECTROMETER SETTING         MWA
           SETTING DETERMINATION OF A THREE AXIS SPECTROMETER BY SIMULTANEOUS STEPS
           SCANNING OF SEVERAL VARIABLES IN NEUTRON INELASTIC DIFFUSION . THE SIX
           ANGLES ARE PUNCHED ON CARDS TO BE USED DIRECTLY BY THE COMMAND UNIT
  314 FORTRAN2 SPC LP    SHOEMAKER          MIXG2/GEN HKL,SPECTROGON STGS,LP,ETC  LPA
           EXECUTIVE PROGRAM AND PACKAGE OF SUBROUTINES. SUBROUTINE INDEX
           GENERATES NEW HKL AND CARTESIAN COORDS OF RL POINT EACH TIME CALLED.
           GENERATES INDICES ASYMMETRIC UNIT ANY LATTICE ANY LAUE GROUP. INPUT
           IS CELL CONSTANTS TRICLINIC OR OTHER.  PGM CALCULATES SETTINGS PHI CHI
           PSI FOR GE XRD5 SINGLE CRYSTAL ORIENTER (GONIOSTAT) OR MAY BE ADAPTED
           TO SETTINGS IN WEISSENBERG OR OTHER GEOMETRY.   OPTIONALLY CALCULATES
           LP OR OTHER NEEDED DATA. COMPILED ON 32K 709, SHOULD RUN ON 8K.
           GENERALIZATION OF PGM 119.
  522 FORTRAN2 SPC       KING M V           DSA2/GONIO STGS FOR CUBIC XTLS        LWA
           PROGRAM COMPUTES SETTINGS CHI, PHI, 2 THETA ON DIFFRACTOMETER EQUIPPED
           WITH GONIOSTAT FOR LAUE-INDEPENDENT REFLECTIONS OF A CUBIC CRYSTAL WITH
           GIVEN LATTICE PARAMETER AND WAVE LENGTH, HAVING ANY GIVEN R. L. VECTOR IN
           VERTICAL (CHI = 90) POSITION.  AVAILABLE IN FORTRAN II AND IN BINARY FOR
           IBM 7094
  464 FTN 1410 SPC       SCHAPIRO           SCO1/PHI,CHI,TWO THETAS FOR XRD-5     LWA
           DATA GENERATED IN LINES WITHIN PLANES ALL OF WHICH INTERSECT IN ONE LINE
  520 FORTRAN2 SPC LS    HAMILTON           MODE1/GONIOSTAT SETTING AND LS.       LWA
           CALCULATES GONIOSTAT ANGLES AND PERFORMS LEAST SQUARES ON OBSRVED
           ANGLES TO REFINE CELL CONSTANTS, ORIENTATION MATRIX, AND SCALE
           ZEROES FOR FOUR CIRCLES.  PHI,CHI, AND TWO THETA CALCULATED FOR FIXED
           VALUE OF OMEGA WHICH NEED NOT BE ZERO.   TESTED ON IBM7094.
 7531 FORTRAN4 SPEC      ALBA,DOME*VAC      PESI/WEIGHTS FOR LS REFINEMENT        LNA
           THE PROGRAM ASSIGNS WEIGHTS FOR LEAST SQUARES REFINEMENT, AND PREPARES
           INPUT TAPE FOR MNQF, MNQC, MNQA. SEVERAL WEIGHTING SCHEMES AVAILABLE.
           TESTED ON 16K 7040.
  505 FORTRAN  SPEC      BAUR               MADELUNG CONSTANTS                    LNA
           EWALD METHOD, ANY SYMMETRY, REPULSIVE AND VAN DER WAALS ENERGY(INVERSE
           POWER). CONTAINS SET OF SOLID ANALYTIC GEOMETRY SUBROUTINES FOR SYSTEMA
           TIC VARIATION OF ATOMIC POSITIONS.
 5059FORTRAN            R.HUBER             3DIM.CONVOLUTION MOLECULE METHOD      LWA
           THE PROGRAMM USES THE CONVOLUTION MOLECULE METHOD (W. HOPPE A.C. 10(1957)
           750, R.HUBER A.C. (1965) IN THE PRESS )TO DETERMINE THE ROTATIONAL AND
           TRANSLATIONAL PARAMETERS OF A KNOWN PART OF A MOLECULE. THE PRGRAMM
           CONSISTS OF 3 PARTS 1.COMPUTATION OF THE STRUCTUREFACTORS OF THE
           CONVOLUTION MOLECULE 2. FOURIERSYNTHESIS 3.SEARCH FOR THE CORRECT
           MOLECULE PARAMETERS.  WITH THE USE OF DATACARDS AND SUBROUTINES THE
           PROGRAMM WORKS WITH ALL SPACEGROUPS IN THE TRICLINIC, MONOCLINIC
           AND ORTHOROMBIC SYSTEMS)
 7029 FORTRAN  SPEC      TOURNARIE          344E/MEAN MOLECULAR PLANE             MPA
  500 FORTRAN2 SPEC      WILLIAMS,D.E.      PACK2/MOLECULAR PACKING ANALYSIS      LWA
           CALCULATES AND MINIMIZES PACKING ENERGY AS FUNCTION OF MOLECULAR
           ORIENTATION, INTERNAL MOLECULAR ROTATION, AND LATTICE CONSTANTS BY
           STEEPEST DESCENT METHOD. ANY SPACE GROUP. CHECKED ON 20K IBM 7074.
 7540 FORTRAN  STAT      DAMIANI,GIGLIO*    DET SYM CENT,RAMACHANDRAN,SRINIVASAN  LNA
```

```
7539 FORTRAN    STAT      DAMIANI,GIGLIO*   DET SYM CENT,HOWELLS,PHILLIPS,ROGERS  LNA
7538 FORTRAN    STAT      DAMIANI,GIGLIO*   DETECTION SYMMETRY CENTRE,WILSON      LNA
 468 FTN 1410   STAT      SCHAPIRO          STAT1,2,3,4/                          LWA
```
CALC. SCL AND DEBYE,C.D.F.,MEAN, AND VAR. OF E**2 FOR SUITABLE SUBSETS
OF DATA. CALC. E,E**2,U,AND PLOT VS HKL
```
 496 FORTRAN    STAT      YAKEL,H.L.        ORSTAT,ORNL-TM-750/WILSON+HPR TESTS   LWA
```
WILSON AVERAGING AND/OR STATISTICAL(HPR) TESTS FOR UP TO 3000 REFLECTIONS
FORTRAN PROGRAM COMPILED AND TESTED ON IBM 7090 AND CDC 1604. AVAILABLE
IN FORTRAN OR BINARY DECKS. 1604 TIME IS APPROX .5 SEC/REFL.
```
7031 FORTRAN4   THETA     JAMARD            BRAGG/GENERATOR FOR HKL,THETA,D,ETC.  LWS
```
GIVEN UNIT CELL (ANY SYSTEM) AND WAVELENGTH, GENERATES HKL, THETA, D,
Q(HKL) = U/D**2, IN ORDER OF INCREASING THETA. TAKES INTO ACCOUNT
EXTINCTION RULES. TESTED ON 1107 UNIVAC.
```
 533 FORTRAN2   TF        SRIVASTAVA,SHO*   VIBELL/ ANISO TF                      LPA
```
INVERTS AND DIAGONALIZES THERMAL MATRIX GIVING PRINCIPAL DIRECTIONS
AND VIBRATIONAL AMPLITUDES. ALL NECESSARY INSTRUCTIONS ON COMMENT CARDS
(NO SEPARATE WRITE-UP).
TESTED ON IBM 7090.
```
3042 FORTRAN2   THETA     HALL              CRYSTAL-COUNTER ORIENTOR              LWA
```
HKL, TWO THETA, PHI, SIN THETA AND SIN TWO THETA ARE GENERATED FROM
THE CELL DIMENSIONS

PROGRAMS LISTED UNDER ACTUAL MACHINE DESIGNATIONS

```
3053 ATLAS      2CDIR     DARLOW J V                                              LPA
```
USES STRUCTURE INVARIANTS METHOD WITH PROBABILITIES TO OBTAIN FROM LIST
OF TRIPLE PRODUCTS POSSIBLE SETS OF SIGNS, EACH WITH A NUMBER GIVING THE
ORDER OF PROBABILITY.
```
3052 ATLAS      CONTOUR   DOLLIMORE         INCLUDED IN FR ABOVE                  LWA
3049 ATLAS      FR        DOLLIMORE         SPACEGROUPS 1-74 NOT 43 OR 70         LWA
3070 ATLAS      DP        DIA, DREW         FIFI/ DATA RECUD. WEISSENBERG
```
PART OF SYSTEM ATSYS (Q. V.)
STORES OUTPUT ON MAGNETIC TAPE.
```
3071 ATLAS      DP        DOL, DREW         PLIP/ DATA REDUC. PRECESSION
```
PART OF SYSTEM ATSYS (Q. V.)
STORES OUTPUT ON MAGNETIC TAPE.
```
3075 ATLAS      DP        DREW              KIKI/STORES DATA MAG TAPE
```
PART OF SYSTEM ATSYS (Q.V.)
STORES DATA ON MAG TAPE DIRECTLY FROM PLANES LIST ON 5-HOLE TAPE. WILL
ALSO OUTPUT DATA ON CARDS FOR BUSING-LEVY 7090 LS AND THE X-RAY 63 SYSTEM
```
3067 ATLAS      DP        DREW              MANIPULATION OF UP TO 3 DATA LISTS
```
ACCEPT UP TO THREE LPID DATA LIST
COMPARES COMMON REFLECTIONS, CORRELATES OR CALCULATES SCALES AND
APPLIES THEM
```
3073 ATLAS      DP        DREW              POLO/ SCALE ON DIFFERENT AXES
```
PART OF SYSTEM ATSYS (Q. V.)
SCRUTINIZES AND SCALES DATA ON TWO OR MORE AXES, TO FORM COMPLETE PLANES
LIST. INCLUDES LS PROCEDURE FOR SCALING.
```
3059 ATLAS F    DP        MOORE,SHEFTER     DP WITH INTERLAYER SCALING            NNN
3074 ATLAS      FR        HARD, DREW        PEPE/ FOURIER
```
PART OF SYSTEM ATSYS (Q. V.)
UP TO MONOCLINIC AND MANY ORTHORHOMBIC, THE REST EASIOY ADDED.
```
3058 ATLAS F    FR SF C   MOORE             3D FOURIER WITH CENTRIC SF CALC       NNN
3077 ATLAS      ID        DIA, DREW         ELSI/ DISTANCE AND ANGLE PGM
```
PART OF SYSTEM ATSYS (Q.V.)
CALCULATES BOND DISTANCES AND ANGLES, ALSO STANDARD DEVIATIONS.
```
3051 ATLAS      ID        DOLLIMORE         INCLUDES BOND ANGLES                  LWA
3066 ATLAS      ID E      DREW              DIST,ANGLS,S.D OF BOTH,ANY SYM,TRANS  LWA
3063 ATLAS F    ID        JMS,SHFTER,MORE   LS PLANE INCLUDED                     NNN
3072 ATLAS      LC        DIAMAND           LSCD/ LS FIT POWD OR S.C. DATA
```
PART OF SYSTEM ATSYS (Q. V.)
CALCULATES LEAST SQUARES FIT ON POWDER OR SINGLE CRYSTAL DATA TO GIVE
ACCURATE CELL DIMENSIONS. APPLICABLE TO ALL CRYSTAL CLASSES.
```
3056 FORTRAN    LS E      FARRAR            FOR1/ HEX TET COHN LS ERROR TEST      LPA
```
USES DIFFRACTOMETER, FILM DATA. CALCULATES LS USING COHEN LS METHOD WITH

WEIGHTING FACTOR, PERFORMS ERROR TEST ON EACH LINE. REJECTS ON
INSTRUCTION. REPEATS LS ANALYSIS. UP TO 50 DIFFRACTION LINES SUITABLE
FOR HEX OR TET. COMPILED ON 1620 60K AUTO-DIVIDE.

```
3057 ATLAS F   LS LC     BRACHER          FORTRAN INPUT HKL AND THETA       LDA
3064 ATLAS     LS LC     DIAMAND          HKL PLUS FUNCTION OF THETA        LWA
3076 ATLAS     LS SF     DIA, DREW        BABA/ BLOCK DIAG LS
```
PART OF SYSTEM ATSYS (Q. V.)
SIMILAR TO ROLLETT MERCURY PROGRAM. BLOCK DIAG., MIXED ISOTROPIC AND
ANISOTROPIC REFINEMENT.
```
3061 ATLAS F   PATSUP    MOORE            BUERGER MINIMUM FUNCTION          NNN
3078 ATLAS     PLANE     SPARKS, HUNT     DIDO/ LS PLANE THRU GROUP OF ATOMS
```
PART OF SYSTEM ATSYS (Q.V.)
CALCULATES WEIGHTED LEAST-SQUARES PLANE THRU GROUP OF ATOMS.
```
3055 ATLAS     ROTN FN   HARDING          ROTATION FUNCTION                 MPA
```
ROSSMANNS FUNCTION, ROTATES WEIGHTED RECIPROCAL LATTICE ON SELF, OR ON
THAT OF SECOND CRYSTAL, AND EVALUATES COINCIDENCE. AUTOCODE.
```
3050 ATLAS     SF        DOLLIMORE        MAY SUBDIVIDE POSITIONS LIST      LWA
3054 ATLAS     SF + FR   HARDING          FOURIER 7.8/WITH STRUCTURE FACTORS MPA
```
3D FR,BL SUMMATION, ANY MONOCLINIC SPACE GROUP, ORTHORHOMBIC ON REQUEST.
MAY BE PRECEDED IN SAME RUN BY MODIFICATION OF F OR BY SF CALCULATION AND
SELECTION OR WEIGHTING OF FO - AGREEMENT ANALYSIS GIVEN. LANGUAGE EMA.
```
3065 ATLAS     SFLS DM   DIAMAND          MIXED REFINEMENT
3060 ATLAS F   SFLS DM   JMS,SHFTER,MORE  BLOCK DIAGONAL                    NNN
4014 ATLAS     2SFRFRDF  DARLOW J V,MAIN,DARLOW S F*/                       LWA
```
CALCS SF. SCALES WITH F OBS USING ALL OR OUTER ONLY REFLEXIONS. GIVES
R FOR ALL AND/OR OUTER TERMS. CALCS FR(OBS) AND/OR FR(CALC) AND/OR DF
USING INNER, OUTER, OR ALL TERMS, ANY OR ALL OF 9 COMBINATIONS POSSIBLE,
UNIQUE AREA ONLY. ANY DIVISION ALONG EACH AXIS. ANY SECTION OF PROGRAM
MAY BE USED BY ITSELF. 2D ONLY.
```
3079 ATLAS     SPEC      DIAMAND          DIAN/ DIHEDRAL ANGLES
```
PART OF SYSTEM ATSYS (Q. V.)
CALCULATES DIHEDRAL ANGLES, ETC. USEFUL FOR MODEL BUILDING.
```
3062 ATLAS F   STAT      MOORE            WILSON PLOT                       NNN
3069 ATLAS     SYSTEM    DREW             ATSYS/ ATLAS CRYSTALLOGRAPHIC SYSTEM
```
SEE UNDER ATSYS (SYSTEMS)
```
 517 B 5000    LS FM     GALLAHER/KAY     TRANSLATION OF ORFLS              LPA
```
TRANSLATION OF PROG.360 TO EXTENDED ALGOL FOR B 5000 (32K)
```
7018 BULL GET LP         BASSI            L AND LP CALC,SIN SQUARE THETA    LWA
```
CALC. L AND LP FACTORS AND SINUS SQUARE THETA
```
7016 BULL GET LS         BASSI            LS.DM.64 PARAMETERS MAX           LWA
```
LEAST SQUARES REFINEMENT, DM, 64 PARAMETERS MAXIMUM. NO TEMPERATURE
FACTOR.
```
7017 BULL GET SF LP      BASSI            SF OR INTENSITIES,LP, N OR XRAYS  LWA
```
SF OR INTENSITIES CALCULATIONS, N OR X RAYS, L OR LP FOR POWDERS,
ALMOST ALL SPACE GROUPS. 16 ATOMIC POSITIONS MAX, 4096 ATOMS PER UNIT
CELL MAX. LF PEPINSKY APPROX.
```
7015 BULLGAET D LP LF    LAB.CALC.CNRS    D,LP(WEIS,P...),LF                LWA
```
TRICLINIC. CALCULATES 1/D, L, AND LP FACTORS (N OR X RAYS)
FOR POWDERS, WEISSENBERG..... CALCULATES LF LINEAR INTERPOLATION
```
5016 BULLGAET DP EXT     IITAKA           EXTINCTION COR FOR STRONG REFLECTION LWA
```
CORRECT F FOR EXTINCTION ASSUMING DARWINS EQ.I*=I/1'GI.WHERE I IS
I OBS. G IS FOUND FOR THE GIVEN STRONG REFLECTIONS AND CALCULATES I* AND
F FOR THE PLANES.
```
5017 BULLGAET DP TF      IITAKA           WILSON PLOT FINDS SCL AND TF      LWA
```
GIVES DATA FOR WILSON PLOT. TF AND SCL FAC. IS CALCULATED BY LS IF THE
WEIGHT FOR EACH ZONE IS ASSIGNED.
```
7013 BULLGAET FR 2       LAB.CALC.CNRS    FOURIER SYNTHESIS 2 DIM.          LWA
```
FOURIER SYNTHESIS 2 DIMENSIONS. STEPS FROM 1/100
```
5019 BULLGAET H 2THETA   IITAKA           LIST H,2THETA ETC IN A GIVEN REGION LWA
```
LISTS HKL,MULTIPLICITY,SIN THETA BY LAMDASQ.,2THETA ETC FOR THE
PLANES WITHIN A GIVEN 2THETA MAX. HKL IS IN A DEFINITE ORDER EXCL.ABSENT
REFLECTIONS.CALCULATION ONLY FOR ANY GIVEN ZONE OR A LAYER OR AN AXIS
IS POSSIBLE.
```
5023 BULLGAET ID         IITAKA           ID BET GIVEN ATOMS ALSO BOND SEARCH LWA
```
ANY SYM.OPERATION INCLUDING LATTICE TRANSLATIONS,DEFINED FOR PARTICULAR
SPACE GROUP IS APPLIED FOR ANY ATOM (IN ASYM UNIT). DISTANCES AND

```
           ANGLES BETWEEN ANY SPECIFIED (OR NOT SPECIFIED) ATOMS ARE CALCULATED.
           IN ANGLE CALCULATION, IT IS POSSIBLE TO FIX ONE ATOM AS CENTER AND
           CALCULATE ALL POSSIBLE DISTANCES AND ANGLES AROUND IT.
5021 BULLGAET LC         IITAKA            FOR ALL XTL SYS BY LS INCL 3 SYST ER LWA
           FINDS RECIPROCAL LATTICE CONST.(MAX 6 PARAMETERS) BY LS METHOD BASED ON
           INDEXED 2THETA OBS BY POWDER OR SINGLE TAKEN BY ANY WAVELENGTH.
           INDIVIDUAL WEIGHT FOR EACH OBS CAN BE PUT. SYSTEMATIC ERRORS UP TO 3
           KINDS ARE INCLUDED. HKL, 2THETA OBS AND CAL TABLE IS PRINTED.
5020 BULLGAET LC TRANS IITAKA              TRANSFORM LC IN BOTH DIRECTIONS     LWA
           GIVES BOTH CELL VOLUMES AT THE SAME TIME.
5018 BULLGAET LP,ABS    IITAKA             WEIS PREC FOR ANY LAYER,XTL SETTING LWA
           WEIS., RETIGRAPH,OSC AND PREC. PHOTO CAN BE TREATED FOR ANY METHOD OF
           TAKING PHOTO,ANY CRYSTAL SETTING (AXIS) AND FOR ANY LAYER LINE.
           ABSORPTION CORRECTION BY INTERPOLATING THE STORED TABLE AND CAN BE
           APPLIED FOR SPHERICAL AND CYLINDRICAL SPECIMENS.
5015 BULLGAET LS 3       IITAKA            TRI,MONO,ORTH,C,DM,ISOTF.EACH ATOM  LWA
           REFINES FOR F SQ.ALL DATA ARE STORED IN THE MACHINE.USES WEIGHTS BY
           HUGHES SYSTEM. THE WEIGHT CAN BE MODIFIED BY A FACTOR GIVEN BY EACH INPUT
           PLANE CARD. THE SHIFTS XYZB FOR EACH ATOM AND OVERALL SCALE FACTOR ARE
           OBTAINED BY DIAGONAL APPROX.THEY ARE MULTIPLIED BY GIVEN DAMPING FAC.
           (XYZ AND B SCL SEPARATELY). AT THE END OF EACH CYCLE SHIFTS,E.S.D.S,
           R ETC ARE LISTED. SPEED FOR TRI. 0.5N+0.28J+1.5 SEC/PLANE.N ATOM J KIND.
5024 BULLGAET PERP. D IITAKA               FIND BEST PLANE,CALC. PERPENDIC. D  LWA
           TRICLINIC CASE. FINDS THE BEST PLANE THROUGH A GIVEN SET OF ATOMS BY
           LS METHOD AND CALCULATES THE PERPENDICULAR DISTANCES FROM ANY GIVEN ATOM.
           THE COEFFICIENTS OF THE EQUATION OF THE BEST PLANE ARE GIVEN WHICH ARE
           PROPORTIONAL TO THE MILLER INDICES OF THE PLANE.
7014 BULLGAET SF 2 3    LAB.CALC.CNRS      SF DIRECT SUMMATION                 LWA
           SF BY DIRECT SUMMATION, GIVEN COORDINATES OF EACH ATOM, HKL SET ,
           AND LF.  1000 ATOMS, MAX VALUE OF H (OR K OR L) 100
5022 BULLGAET SF 3 FR2 IITAKA              TRI MONO ORTH C FOLLOWED BY FR PROJ LWA
           ISO. TF FOR EACH ATOM KIND. DATA FOR ANY FR OR DIF FR PROJ. ARE STORED
           DURING SF CAL. BUT SF INPUT IS FOR UNIQUE REF. R FAC. ETC ARE LISTED.
           FR CAL. CAN START AFTER SF FOR ANY PART OF THE CELL WITH ANY INTERVAL.
           OUT PUT PERMITS DIRECT CONTOURING. SF SPEED FOR TRI. 0.06N+0.52J+3 SEC/
           PLANE. N ATOM J ATOM KIND.
7012 CAB 500    ID         CNRS BELLEVUE   ID,BOND ANGLES,TRICLINIC,ORTHORHOMB.LWA
           INTERATOMIC DISTANCES AND BOND ANGLES FOR TRIPLETS OF ATOMS IN
           TRICLINIC AND ORTHORHOMBIC SYSTEMS
7004 CAB 500   LF 1/D2 CNRS BELLEVUE     CALC LF,1/D SQUARE ALL SYSTEMS        LWA
           CALCULATES 1/D SQUARE AND LF, ALL SYSTEMS, HKL SET AND LF VALUES FOR
           DISCRETE VALUES OF 1/D BEING GIVEN. LF BY LINEAR INTERPOLATION
7011 CAB 500   LP       CNRS BELLEVUE     LP CALC,WEIS EQUI-INCL.H.ALL SYSTEMS LWA
           IN ALL SYSTEMS, GENERATES HKL LIMITED BY REFLECTION SPHERE. FOR EACH
           HKL, GIVES 1/D SQUARE , AND LP FOR WEIS EQUI-INCL.
7002 CAB 500   LS 2     CNRS BELLEVUE     LS 2 FOR DIFFERENT PLANE GROUPS.DM   LWA
           ONE PGM OF LS REFINEMENT FOR EACH OF PLANE GROUPS P2, PGG, PMG, CMM,
           PMM.  REFINES ATOMIC COORDINATES, ISOTROPIC TF (THE SAME FOR ALL ATOMS),
           AND SCALE FACTOR. DM. MAX NUMBER OF ATOMS UNRELATED BY SYMMETRY 20. MAX
           NUMBER OF REFLECTIONS 300. 10 DIFFERENT ELEMENTS MAX.
7001 CAB 500   LS 3     CNRS BELLEVUE     LS 3 FOR P212121 SPACE GROUP.DM      LWA
           LS REFINEMENT FOR P212121 SPACE GROUP, ON THE ATOMIC COORDINATES OF 20
           INDEPENDENT ATOMS MAX.  300 REFLECTIONS MAX.  LFS PROPORTIONAL TO A
           UNIQUE LF. DM.
7006 CAB 500   SCL TF    CNRS BELLEVUE    SCL AND TF WILSON METHOD             LWA
           SCALE AND TEMPERATURE FACTORS BY WILSON METHOD. 1400 VALUES OF
           EXPERIMENTAL INTENSITIES MAXIMUM.
7007 CAB 500   SF 2      CNRS BELLEVUE    CALC SF 2 FROM AT.COORD. OF EACH AT. LWA
           CALCULATES SF 2 FROM ATOMIC COORDINATES OF EACH ATOM, HKL AND CORRES-
           PONDING LFS BEING GIVEN. 40 ATOMS MAX. (OR 80 IF CENTERED).
           4 DIFFERENT ELEMENTS MAX.
7009 CAB 500   SF 2 3 C CNRS BELLEVUE     SF 2 3 TRICL C WITH ISOTROPIC TF     LWA
           CALCULATES SF 2 OR 3 FOR TRICLINIC CENTERED CELL. ISOTROPIC TEMPERATURE
           FACTOR IF NEEDED.  HKL SET, 1/D SQUARE, AND LFS ARE GIVEN FROM ANOTHER
           PROGRAM.
7008 CAB 500   SF 3      CNRS BELLEVUE    CALC SF 3 FROM AT.COORD. OF EACH AT. LWA
```

```
       CALCULATES SF 3 FROM ATOMIC COORDINATES OF EACH ATOM, HKL AND CORRES-
       PONDING LFS BEING GIVEN.  100 ATOMS MAX. (OR 200 IF CENTERED).
       4 DIFFERENT ELEMENTS MAX.
7010 CAB 500   SF 3 C    CNRS BELLEVUE    SF 3C WITH INDIVIDUAL ISOTROPIC TF   LWA
       CALCULATES SF 3 FOR CENTERED CELL. ISOTROPIC TEMPERATURE FACTOR
       FOR EACH ATOM, IF NEEDED. HKL SET, 1/D SQUARE, AND LFS ARE GIVEN FROM
       ANOTHER PROGRAM.
7005 CAB 500   SPEC      CNRS BELLEVUE    F OBS FROM OPTICAL DENSITIES         LWA
       CALCULATION OF F OBSERVED AS SQUARE ROOT OF OPTICAL DENSITIES
       CORRECTED WITH LP FACTORS
7003 CAB 500   SPEC      CNRS BELLEVUE    NORMALIZES INTENSITIES FROM N FILMS  LWA
       PUTS ON THE SAME SCALE INTENSITIES FROM N FILMS. N MAX = 12.  GIVES FOR
       EACH REFLECTION THE N VALUES ON SCALE OF FIRST FILM AND MEAN VALUE.
7526 CEP       SF C AT   GUERRI,BERTOLUZ  MIN SF/                              LWS
7527 CEP       SF C 2    GUERRI,BENVENUT  MIN F/                               LWS
6008 DASK K2   FR 2 3    LARSEN F         DASK 330D/GENL FR 3 DIM              LWA
       3-DIM FR PGM APPLICABLE TO ANY SPACE GROUP.
       STRAIGHT OR DIFFERENCE FR AND BOUNDED PROJECTIONS POSSIBLE.
6009 DASK K2   SF        MONDRUP          DASK 330/GENL SF                     LWA
       SF PGM, ANISOTROPIC TF, USES TRICLINIC FORMULAE.
4023 DEUCE     DP        SIME/SPEAKMAN    LP TUNNELL CORR FOR WEISSENBERG      MWA
       P014/EQUI-INCLINATION WEIS DATA FOR A TRICLINIC CRYSTAL ARE CORRECTED FOR
       LP TUNNELL FACTORS. BASIC DEUCE GIVES 50 PLANES PER MINUTE. OUTPUT
       SUITABLE FOR ROLLETT SF SFLS FR PGMS.
4028 DEUCE     FR        ROLLETT/PETERS   FR/3D BL GENL                        MWA
       ALL SPACE GROUPS BUT MUST SUM OVER ALL REFLECTIONS WITH NON-NEGATIVE
       INDICES.  NO LIMIT ON REFLECTIONS. SECTIONS MUST BE DONE IN PARTS IF MORE
       THAN 60 BY 80 POINTS. INTERVALS N/240.
4021 DEUCE     FR 2      SIME/SPEAKMAN    ELECTRON DENSITY IN GENERAL PLANE    MWA
       P028,029,030/PROGRAM COMPUTES FOURIER SECTION AT A GENERAL PLANE THRU A
       MONOCLINIC CELL USING THE METHOD OF TREUTING AND ABRAHAMS ACTA CRYST. 14,
       190.
4049 DEUCE     FR 3 DF   SIME/SPEAKMAN    DIFFERENTIAL SYNTH                   NPX
4020 DEUCE     ID        SIME/SPEAKMAN    BOND LENGTH AND ANGLE                MWA
       P031,032/BOND LENGTHS AND ANGLES CALCULATED FOR UP TO 128 ATOMS AND CAN
       GENERATE FROM THESE TWELVE RELATED SETS OF 128 ATOMS.
4025 DEUCE     PLANE     WATSON/SPEAKMAN  MEAN MOLECULAR PLANE DETERMINATION   MPX
       P069/ALPHACODE PROGRAM DETERMINES BEST PLANE THROUGH A SET OF ATOMS BY
       METHOD OF SCHOMAKER, ACTA CRYST. 12, 600. ATOMS CAN BE GIVEN ANY WEIGHT
       FOR THE LEAST SQUARES PROCESS. DISTANCES OF ALL ATOMS FROM THIS PLANE ARE
       ALSO OBTAINED.
4022 DEUCE     R SCL     SIME/SPEAKMAN    SCALE AND DISCREPANCY FACTOR         MWA
       P019/BASIC DEUCE, SCL AND R FROM ROLLETT STYLE SF OUTPUT.
4029 DEUCE     SF        ROLLETT/PETERS   SF/3D GENL ISOTROPIC                 MWA
       ALL SPACE GROUPS NO LIMIT ON REFLECTIONS. PARAMETERS, F SCALE,
       POSITIONS, ISOTROPIC B FACTORS, CHEMICAL TYPES.  UP TO 1664 ATOMS
       OF 16 TYPES.
4030 DEUCE     SF LS     ROLLETT/PETERS   LS/3D BLOCK DM AT GENL               MWA
       ALL SPACE GROUPS, NO LIMIT ON REFLECTIONS. PARAMETERS, F SCALE,
       POSITIONS, ANISOTROPIC B FACTORS, CHEMICAL TYPES.  UP TO 64 ATOMS
       OF 16 TYPES.
4024 DEUCE     T/OMEGA   SIME/SPEAKMAN    CRUICKSHANK THERMAL ANALYSIS ACTA 9  MWX
       P036,037,038,039/ALPHACODE AND GIP MONOCLINIC THERMAL ANALYSIS PGM. 1)
       TRANSFORMS COORDINATES, TEMP FACTORS TO ORTHOG MOLECULAR AXES. GIVES
       VALUES OF U SUB IJ FOR EACH ATOM (ACODE). 2) FORMS LS NORMAL EQTNS FOR
       DETN OF T AND OMEGA (ACODE). 3) INVERTS MATRIX, SOLVES FOR T AND OMEGA
       (GIP7). 4) DETERMINES STANDARD DEVIATIONS (ACODE).
5073 ER 56     ABS       THURN            ABSORB/WEIS,IRREG POLYHEDR CRYSTAL   LWA
       IT LC, EQU OF CRYST LIMITING PLANES, SF, OT CORRECTED SF
5007 ER 56     FR 2      THURN            GENL, VARIABLE GRID, H, K UP TO 19   LWA
       NEEDED FOR IT -- ABSOL VALUES OF STRUCT AMPLITUDES, PHASE ANGLES
5004 ER 56     FR 2 C    KRUECKEBG,BERG.  2DIM ALL CENTROSYM SPACE GROUPS      LPA
5005 ER 56     ID        MATTES           DIST. ANGLES BETWEEN SELECTED ATOMS  LWA
       ANY SYMMETRY. DISTANCES AND BOND ANGLES BETWEEN SELECTED ATOMS
5003 ER 56     LC BY LS  SCHMID           STRUCTUR E/ REFINEMENT OF LC BY LS   LWA
       GILT FUER ALLE KRISTALLKLASSEN. IT. WELLENLAENGE UND (UNGENAUE
```

LC ODER KOEFFIZIENTEN DER QUADRATISCHEN FORM) SOWIE (THETA ODER
4XTHETA ODER SINUSQUADRAT THETA) UND KENNZIFFER FUER KRISTALLKLASSE.
OT. VERBESSERTE KOEFF. DER QUADR. FORM, E, LC VERBESSERT, (SINQUADRAT
THETA FUER GEGEBENE (HKL)) GEMESSEN UND MIT NEUEM LC BERECHNET.
BIS 200 THETAWERTE ZUGELASSEN.

```
5006 ER 56      LP WEIS MATTES         WEIS LP FOR ANY LEVEL              LWA
```
CALC OF SF, SQUARE OF SF, SIN THETA FROM EQUIINCLINATION
WEIS'DATA
```
5001 ER 56      THETA+SF SCHMID        STRUC. A/IN MILLER INDICES ARRANGED LWA
```
SINQUADRAT THETA UND KOMPLEXER SF FUER ALLE KRISTALLSYSTEME 3'DIM.
BIS ZU 200 ATOME AUS 10 VERSCHIEDENEN ATOMSORTEN JE ELEMENTARZELLE
ZUGELASSEN. AUSGABE IN LEXIGRAPHISCHER ANORDNUNG DER MILLER INDICES
```
5002 ER 56      THETA+SF SCHMID        STRUCTUR B/ARRANGED INCREASING THETA LWA
```
FUER MONOKLINE UND HOEHERE SYMMETRIE ERHAELT MAN SINQUADRAT THETA,
THETA UND KOMPLEXEN SF, DER GROESSE VON THETA NACH GEORDNET. BIS ZU
200 ATOME AUS 10 VERSCHIEDENEN ATOMSORTEN JE ELEMENTARZELLE SIND
ZUGELASSEN. ERFASST WERDEN DIE MILLERINDICES H='9(1)+9,
K=0(1)9, L=0(1)9.
```
6020 FACIT 2K ABS     WERNER          ABSP/PREC 3, XTALS OF ARBITR SHAPE   LWA
```
AS ABSW FOR FACIT EXCEPT THAT CALC TIME IS C. N*M**3*0.018 SEC PER REFL
```
6019 FACIT 2K ABS     WERNER          ABSW/WEISS 3, XTALS OF ARBITR SHAPE  LWA
```
CALULATES ABS FACTORS AND CORRECTS INTENSITIES. GAUSS METHOD FOR
NUMERICAL INTEGRATION. N (MAX. 14) XTAL SURFACES. M**3 VOLUME ELEMENTS
TO BE CHOSEN SO THAT M**3* (N+1) IS LESS THAN 8000. INPUT IS EQN. FOR -
OR COORDINATES FOR THREE POINTS IN - EACH XTAL SURFACE, CELL DIMENSIONS,
LINEAR ABSORPTION COEFFICIENT, WAVE LENGTH, HKL AND OBSERVED
INTENSITIES. ABOUT N*M**3*0.009 SEC PER REFLECTION.
```
6029 FACIT 2K ABS-EXT WERNER          ABSG/SEC EXTINCTION AND ABS          MPA
```
GE SINGLE XTAL DIFFR DATA. AS ABSP FOR FACIT EXCEPT THAT CALCS. FOR SEC
EXTINCTION CORR ARE INCLUDED. ABSG USES RESULT FROM PGM GALPI FOR FACIT
```
6034 FACITAGL DP      NORRESTAM       AUTO-GALPI/LP,WEIGHTS ETC.           MPA
```
PGM GALPI MODIFIED FOR HANDLING RESULT FROM G.E. AUTOM. SINGLE XTAL DIFF.
```
6033 FACITAGL DP      NORRESTAM       GALPI/LP, WEIGHTS ETC.               LWA
```
PGM CALCULATES LP, STATISTICAL WEIGHTS AND DIRECTION COSINES OF X-RAYS
FOR G.E. MANUAL SINGLE X-TAL DIFFRACTOMETER RESULT. ANY SYMMETRY. 2THETA-
THETA SCAN. RESULT USED BY PGM ABSG FOR FACIT
```
6030 FACITAGL E       ASB/NORRESTAM   STINTA/E IN ID                       LPA
```
PGM CALCULATES E IN ID FROM E IN COORDS AND LC FOR ANY SYMMETRY. ALL
ATOMS ARE TREATED AS INDEPENDENT. USES RESULT FROM MODIFIED PGM VINTER-2
```
6014 FACIT    FR 2 3  LIMINGA,OLOVSON PROFFS/ANY SYMM,GRID.REFL ANY ORDER. LWA
6016 FACIT    ID      LIMINGA,OLOVSON VINTER/DIST,ANGLES.ANY SYMMETRY.     LWA
6026 FACIT 2K LF      WESTMAN         AFFE/LS FIT POLYNOM(SIN THETA)=LOG LF MPA
6017 FACIT    LP      WERNER          PREC, LP FOR ANY LEVEL, H            LWA
6024 FACIT 2K LP      WESTMAN,ASBRINK LOPPA/LP OF INTENSITIES FROM WEIS    MPA
6023 FACIT    LS      ASBRINK,BRANDEN LS/BLOCK-DM APPR,LAYER SCLS,ISOTR TFS LWA
```
SFLS (OR SF) PGM FOR ORTHORH SYMM AND LOWER. PGM REFINES F BY SHIFTING
LAYER SCALEFACTORS COUPLED IN A MATRIX WITH AVERAGE VIBRATION PARAMTR,
AND BY SHIFTING ATOMIC COORDS AND INDIV ISOTR VIBR PARAMTRS USING BLOCK
DIAG MATRIX APPROXIMATION. PGM USES LF TABLES, PROVIDES THREE DIFFERENT
WEIGHTING SCHEMES, HAS ACCELERATION DEVICES AND MAKES WEIGHT ANALYSIS.
ACCIDENTALLY ABSENT REFLS, IF INCLUDED, ARE TREATED SEPARATELY.
```
6027 FACIT 2K P       WERNER          PR1/PI APPR KNOWN PTRN, REFINES LC   LWA
6028 FACIT 2K P       WERNER          PR2/PI TRIAL-AND-ERROR,ORTHORH SYMM  LWA
6015 FACIT    SF H    LIMINGA,OLOVSON STRIX/CALC SF+H FOR FR IT.ANY SYMM.  LWA
6035 FACIT    SFLSFMCS LUNDBERG,B.K.S. ROHAP/REF.OF PARAM.ISOM.REPL. PROT.  LPA
```
METHOD DESCRIBED IN ACTA CRYST. (1965) 18, 576. SOON IN FORTRAN.
```
6032 FACITAGL SPC     NORRESTAM       AUTO-GIP/SPC GONIOSTAT               MPA
```
PGM GIP MODIFIED FOR STEERING G.E. AUTOMATIC SINGLE XTAL DIFFRACTOMETER
USING 2THETA-THETA SCAN. SCAN INTERVAL CALCULATED AS FUNCTION OF THETA
```
6031 FACITAGL SPC     NORRESTAM       GIP/SPC GONIOSTAT                    LWA
```
SPC FOR G.E. MANUAL SINGLE XTAL DIFFRACTOMETER. ANY SYMMETRY. H IN
DESIRED OCTANT. EXTINCTION CONDITIONS MAY BE SPECIFIED
```
6513 LGP30    D       LINEK*NKB       FOR THETA INTERPOLATE IN TABLES      NPA
```
INPUT ARBITRARY TABLES, LATTICE CONSTANTS. OUTPUT VALUES OF
PHYSICAL FUNCTIONS FROM TABLES FOR EVERY HKL.
```
   1 LGP30    DP      KOENIG          M-13 100/AVG WEISS OR PREC DATA, LP  LWA
```

```
   3 LGP30    FR 2 A    KOENIG              M-13 102/BL 1/64 ACENTR           LWA
   2 LGP30    FR 2 C    KOENIG              M-13 101/BL 1/64 CENTRO           LWA
6512 LGP30    ID        LINEK*NKB           GENERAL FORMULA                   LWA
3514 KDF9 AUT A-C TEST  HAMOR               ACENTRIC-CENTRIC STATISTICAL TEST LWA
     STATISTICAL TEST FOR CENTER OF SYMMETRY BY METHOD OF HOWELLS, PHILLIPS,
     AND ROGERS. TREATS HKL,HK0,H0L,0KL DATA SEPARATELY. AXIAL REFLECTIONS
     OMITTED. INPUT AS FOR ROLLETT SF LS PROGRAM.
3516 KDF9 AUT DP        HAMOR,WATKIN        LP TUNNELL CORR FOR WEIS DATA     LWA
     OT SUITABLE FOR ROLLETT SF LS PROGRAM. PROGRAM ALSO AVAILABLE TO
     CONVERT THIS OT TO FORM SUITABLE FOR CRUICKSHANK,SIME ALGOL PROGRAMS.
3517 KDF9 AGL FR 3      SIME                GEN FR ALL SG INCL CONTOUR OT     MPA
3518 KDF9 AGL ID        MUIR,MCGREGOR                                         MPA
3515 KDF9 AUT ID        WATKIN,HAMOR        INTRA-,INTER-MOLEC DIST AND ANGLES LPA
     GIVEN ATOMIC COORDS AND EQUIVALENT POSITIONS (TRICLINIC,MONO, OR ORTHO)
     CALCULATES ALL INTRAMOLECULAR DISTANCES BELOW GIVEN LIMIT, AND INTER-
     MOLEC DISTANCES BELOW SECOND GIVEN LIMIT. ANGLE CALC. DETERMINED BY
     SAME LIMITS.
3513 KDF9 AUT PLANE     HAMOR               BEST PLANE THROUGH SET OF POINTS  LWA
     EQUATION OF PLANE THRU SPECIFIED POINTS SELECTED FROM INPUT COORDS
     CALCULATED BY METHOD OF SCHOMAKER ET AL, AND BLOW. OT INCLUDES
     DEVIATIONS OF SPECIFIED POINTS FROM PLANE, MEAN AND RMS DEV AND DIST
     OF OTHER POINTS FROM PLANE.
3512 KDF9 AGL SFLS      CRUICK,SMITH JG     ALL SG,AT,DISPERSIN,SCLS,DM OR FM,E MPA
4011 MERCURY  DIRC      WOOLFSON            WOOLFSONS FOLLY/                  LPA
     DATA - SIGN RELS IN FORM OF TRIPLE PRODUCTS.
     OUTPUT - ALL SETS OF SIGNS FOR WHICH SIGN RELS ARE VALID WITHIN
              PRESCRIBED LIMITS.
     METHOD - AS DESCRIBED IN ACTA CRYST 10, 116.
     TIME - DEPENDS ON PRESCRIBED LIMITS. (SEE ALSO PGM 4002, FOR 704.)
4031 MERCURY  DP        LAI/ROLLETT         CORRELATE F-SQUARED               LPA
     GIVEN F-SQUARED FROM 2 SETS INTERSECTING LAYERS AND SCL FACTORS,
     PRODUCES AVERAGED DATA ON COMMON SCALE.
4038 MERCURY  DP        MAIR/ROLLETT        CORRELATE AND SORT                MWA
     GIVEN F-SQUARED FROM 2 SETS INTERSECTING LAYERS AND SCL FACTORS,
     PRODUCES AVERAGED DATA ON COMMON SCALE. ANSWERS CAN BE SORTED BEFORE
     OUTPUT.
4032 MERCURY  DP        ROLLETT             SORT HKL DATA                     MPA
     GIVEN HKL DATA IN RANDON ORDER, PROVIDES DATA SORTED INTO ANY DESIRED
     ORDER OF INDICES.
4039 MERCURY  E         MAIR/ROLLETT        BOND LENGTHS AND STANDARD DEVIATIONS
     GIVEN POSITION AND VARIANCE MATRIX OUTPUT FROM 3-D LS, FINDS BOND LENGTHS
     AND THEIR STANDARD DEVIATIONS.
6002 MERCURY  FR 2      KEILHAU             NDRE-R2280/COSCOS OR SINSIN VERSION LWA
     COMPUTES 2-DIM FR AS A SUMMATION OVER F (HKL) COSCOS-
     OR F (HKL) SINSIN-TERMS ONLY. SPEC CASE OF PGM NDRE-R2281.
     AXES SUBDIVISION 1/60. INPUT PGM NDRE BIP.
6003 MERCURY  FR 2      KEILHAU             NDRE-R2281/COSCOS + OR - SINSIN VERSN LWA
     COMPUTES 2-DIM FR AS A SUMMATION OVER (A(HKL) COSCOS + OR -
     B (HKL) SINSIN) TERMS. AXES SUBDIVISION 1/60. INPUT PGM NDRE BIP.
6004 MERCURY  FR 2      KEILHAU             NDRE-R2282/PLANE GROUPS 7,8 AND 12 LWA
     COMPUTES 2-DIM FR AS A SUMMATION OVER A(HKL) COSCOS TERMS WHEN Q IS EVEN,
     B (HKL) SINSIN TERMS WHEN Q IS ODD, Q BEING H OR K OR H+K.
     AXES SUBDIVISION 1/60. INPUT PGM NDRE BIP.
6005 MERCURY  FR 2      KEILHAU             NDRE-R2284/GENERAL VERSION        LWA
     COMPUTES 2-DIM FR AS A SUMMATION OVER TERMS A (HKL) COSCOS+B (HKL) SINCOS
     +C (HKL) COSSIN-D(HKL) SINSIN. AXES SUBDIVISION 1/60. INPUT PGM NDRE BIP.
6006 MERCURY  FR 3      KEILHAU             NDRE-R2285/230 OPTIMAL PGMS       LWA
     CONSTRUCTS OPTIMAL WORKING PGM VALID FOR THE SPACE GROUP IN QUESTION
     TAKING FULL ADVANTAGE OF COMPUTATIONAL SHORTCUTS MADE POSSIBLE BY
     SYMMETRY RELATIONS. WORKING PGM IS VERY CLOSE TO ABSOLUTE OPTIMUM.
     FINALLY PERFORMS 3-DIM FR. AXES SUBDIVISION 1/60. INPUT PGM NDRE DIP.
4046 MERCURY  GEOMETRY  ROLLETT             ROTATE BOND                       MPA
     GIVEN POSITIONS OF THREE ATOMS FORMING TWO BONDS, FINDS POSITION BY
     ROTATING ONE BOND ABOUT ANOTHER.
4047 MERCURY  GEOMETRY  SPARKS/ROLLETT      LOCATE HYDROGEN ATOMS             MWA
     THREE ROUTINES WHICH PROVIDE COORDINATES FOR H ATOMS CALCULATED FROM
```

                    OTHER POSITIONS FOR ATOMS JOINED TO XH, XHH, XHHH GROUPS RESPECTIVELY.
4043 MERCURY GEOMETRY SPARKS/ROLLETT   MOLECULAR AXES   MWA
                    GIVEN ATOMIC POSITIONS AND WEIGHTS, FINDS CENTROID, AXES OF INERTIA
                    (BY LATENT VECTOR ANALYSIS) AND COORDINATES REFERRED TO THESE AXES.
4045 MERCURY ID   SPARKS/ROLLETT   DISTANCES AND ANGLES   MWA
                    GIVEN ATOMIC POSITIONS AND SYMMETRY OPERATIONS, PROVIDES ALL UNIQUE
                    DISTANCES IN STRUCTURE BELOW GIVEN LIMIT, ALL ANGLES WITH LEGS BELOW
                    SECOND LIMIT.
4012 MERCURY LC   MAIN,WOOLFSON   FROM WEIS DATA   LPA
                    USES ALPHA 1 - ALPHA 2 SEPARATION FROM ZERO-LAYER WEIS. INPUT -
                    INDICES WITH MEASURED SEPARATIONS, CAMERA CONSTANTS, ROUGH LC.
                    OUTPUT - ACCURATE LC WITH E. ACCURACY ABOUT 1/2000.
4008 MERCURYP P H D SF PAULING,DOLLIMORE/RD1/CALC,SORT D, SF   NPA
                    FOR GIVEN CELL, GENERATES HKL, CALCULATES D FOR ALL PLANES IN GIVEN
                    SHELL, SORTS IN ORDER OF S, PUNCHES ANY OF D, 1/D, D2, 1/D2, S, S2,
                    LAMBDA S, (LAMBDA S)2, THETA, 2THETA. FOR GIVEN STRUCTURE WILL PRODUCE
                    F, I, ILP. ISOTROPIC TF ONLY, ANY SPACE GROUP.
4040 MERCURY PATSUP   ABRHMSN,ROLLETT   PATTERSON SUPERPOSITION 3D   LPA
                    GIVEN 3-D PATTERSON FUNCTION, SYMMETRY OPERATIONS, TRANSLATES ORIGIN TO
                    EACH OF UP TO 16 GRID POINTS, CALCULATES SPECIFIED FUNCTION OF
                    COINCIDENT POINTS.
4048 MERCURY PK LOCN SPARKS/ROLLETT   19-POINT GAUSSIAN INTERPOLATION   MWA
                    GIVEN 19 POINTS OF FO MAP PROVIDES PEAK POSITION AND HEIGHT. GIVEN FC
                    (OR FO - FC) POINTS AS WELL AND INPUT POSITION CALCULATES N-SHIFTS AND
                    BACK-SHIFTS. GIVES B-FACTOR ALSO IF AXES ORTHOGONAL.
4007 MERCURY LS2 FMAT CURTIS/BOWLER   29F/GENL 2D LS,FM,AT,N OR XRAY   LWA
                    GENERAL 2-DIM SF + LS PGM, FULL MATRIX, ANISO. NEUTRON OR XRAY. INPUT
                    PLANE GP NO, LC, LF, LIST OF HKF, PARAS, CONTROL CODE FOR EACH CYCLE.
                    WEIGHT IS A FUNC OF F. OUTPUT PARAS, CHANGES, ERRORS, SF, AS REQD.
                    WRITE-UP AERE-R3134 AMENDED BY CPN 52. SAME AS 7090 PGM LS2D,NO.
                    UP TO 31 DISTINCT ATOMS, 85 PARAS REFINED, 500 PLANES, ON 24K MERCURY
4041 MERCURY R   LAI/ROLLETT   AGREEMENT ANALYSIS   LPA
                    GIVEN F-OBS, F-CALC, FINDS SUMS OF MODULI OF F-OBS, F-CALC, F-OBS MINUS
                    F-CALC, AND R FACTORS, FOR GROUPS INDICATED BY F-OBS, SIN THETA, AND
                    INDEX VALUES.
4033 MERCURY SCLLAYRS ROLLETT   LAYER SCL FROM RELATIVE F-SQUARED   LPA
                    GIVEN F-SQUARED FROM LAYERS WHICH INTERSECT, FINDS SCL FACTORS TO PLACE
                    LAYERS ON SAME SCL BY LATENT VECTOR ANALYSIS.
4035 MERCURY SF LS   ROLLETT   LS/3D BLOCK DM GENL   MPA
                    ALL SPACE GROUPS. LIMIT OF 6000 REFLECTIONS, PARAMETERS, F SCALE,
                    POSITIONS, ISOTROPIC B FACTORS, CHEMICAL TYPES. UP TO 96 ATOMS OF
                    16 TYPES.
4036 MERCURY SF LS   ROLLETT   LS/3D BLOCK DM AT GENL   MWA
                    ALL SPACE GROUPS. LIMIT OF 6000 REFLECTIONS, PARAMETERS, F SCALE,
                    POSITIONS, ANISOTROPIC B FACTORS, CHEMICAL TYPES. UP TO 96 ATOMS OF
                    16 TYPES.
4042 MERCURY TFANALYS SPARKS/ROLLETT   MOLECULAR VIBN-LIBN TENSORS   MWA
                    GIVEN ATOMIC POSITIONS, TF, AND WEIGHTS, FINDS AXES AND AMPLITUDES OF
                    VIBRATION AND LIBRATION. LIBRATION AXES PASS THROUGH CENTRE OF MASS.
                    OBSERVED AND CALCULATED ATOMIC VIBRATION TENSORS ALSO OUTPUT.
4044 MERCURY TFANALYS SPARKS/ROLLETT   ATOM VIBRATION AXES   MWA
                    GIVEN AT FOR ATOM, PROVIDES PRINCIPAL VIBRATION AXES AND AMPLITUDES BY
                    LATENT VECTOR ANALYSIS.
4013 MERCURY 3ABSLP   WOOLFSON   EQUI-INCLINATION WEIS   LWA
                    LIMITED TO CRYSTALS IN FORM PARALLELEPIPED WITH EDGES ALONG CELL
                    DIRECTIONS. SIMPLE INPUT. OUTPUTS ABS CORRECTION, LP, PHILLIPS SPOT-
                    SHAPE FACTOR, SIN THETA. TIME DEPENDS ON CRYSTAL SIZE - NORMALLY 2 SECS
                    PER REFLECTION.
8518 M1B   FR 2   NIIZEKI   XTL2/GENL,A,C,GRID N/1000   LWA
                    TREATS ANY SPACE GROUP AS TRICLINIC. A OR C BY PGM PARAMETER. SIN TABLE
                    LOOK UP. INPUT IS MATRIX OF F VALUES WITH PROPER END MARKS. GRID N/1000.
8515 M1B   ID   NIIZEKI   DIST,ANGLES,ANY SYM   LWA
                    PGM NEEDS INDICATIONS OF ATOMS BY PROPER NUMBERINGS. COMPUTES INTERATOMIC
                    DISTANCE OF GIVEN PAIR OF NUMBERS OR ANGLE OF GIVEN TRIPLET OF NUMBERS.
8516 M1B   LP ABS   NIIZEKI   XTL3/LP,ABS CORR HEX PRSM XTLS   LPA
                    APPLICABLE TO EQUATOR REFLECTIONS FROM CRYSTAL WITH UP TO 6 SIDED PRISM

```
      SHAPE. INPUT IS INDICES CRYSTAL OUTLINE SIN THETA/LAMDA AND ORIGINAL F
      SQUARED. PGM TAKES ANY NUMBER OF REPRESENTATIVE POINTS AND INTEGRATES
      BEAM PATHS.
8514 M1B       SF 2    NIIZEKI         XTL4/FOR 14 PLANE GROUPS              LPA
      SUBROUTINE GENERATES ATOMIC F VALUE OF EACH ATOM FOR EACH REFLECTION.
      PGM SELECTS ANY ONE OF FOURTEEN PLANE GROUPS BY PGM PARAMETER. USES
      ISOATOMIC TEMPERATURE FACTORS.
8517 M1B       SPC     NIIZEKI         XTL1/GONIO STGS PO CONTROL TAPE       LWA
      NORMAL BEAM ANY LEVEL ANY ROTATION AXIS. INPUT IS CELL CONSTANTS
      WAVELENGTH AND MAXIMUM THETA. PGM GENERATES INDICES CALCULATES SETTINGS
      PHI UPSILON FOR RIGAKUDENKI SXG1 GONIOMETER FOR SINGLE CRYSTAL. ALSO
      PUNCH OUT CONTROL TAPE FOR AUTOMATIC MEASUREMENTS OF INTEGRATED INTENSI-
      TIES.
6529 NE 803B   FR      SASVARI*GERGELY FOR THE 2 DIM SGR P2, PGG, PGM, PMM   LWA
      THE FOURIER PROJECTION OR THE CORRESPONDING PATTERSON FUNCTION WILL BE
      COMPUTED DEPENDING ON A SPECIAL REFERENCE NUMBER GIVEN AMONG THE DATA.
      THE PROGRAMS (WRITTEN IN ACODE) WITH DIRECTIONS OF USE ARE PUBLISHED
      (ENGLISH) IN ACTA CHIM. HUNG., 1964.
6531 NE 803 B  FR      TICHY           2,3 FOR S. G. P2/M, P21/C, C2/C       LWA
6514 NE803     ID      LINEK           GENERAL FORMULA                       LWA
6525 NE 803B   LP      SASVARI*SANTA   TRICL., MONOCL. AND ORTHORHOMB.       LWA
      ALL POSSIBLE VALUES OF (HKL) WILL BE COLLECTED ON THE BASIS OF THE
      SIN**2 THETA VALUES AND THE 1/LP COMPUTED FOR EQUI-INCLINATION WEISS.
      PHOTOGRAPHS WITH ROTATION AROUND ALL THREE AXES. THE PROGRAMS (WRITTEN IN
      AUTOCODE) WITH DIRECTIONS OF USE ARE PUBLISHED (ENGLISH) IN ACTA CHIM.
      HUNG., 1964.
6533 NE 803 B  LS      TOMAN,OCENASKOVA*/C,E,ICA,R,SCL,TF FOR MONOCL.SP.GRPS LWA
6527 NE 803B   SCL     SASVARI         ABSOLUTE AND UNITARY SCALE FACTORS    LWA
      THE DATA OF SCALE FACTORS FOR THE DETERMINATION OF ABSOLUTE AND UNITARY
      STRUCTURE FACTORS WILL BE COMPUTED. THE PROGRAM (WRITTEN IN ACODE) WITH
      DIRECTION OF USE IS PUBLISHED (ENGLISH) IN ACTA CHIM. HUNG., 1964.
6530 NE 803B   SF      SASVARI*SANTA   MONOCLINIC P21/C                      LWA
      THE STRUCTURE FACTORS WITH OR WITHOUT ISOTROPIC TEMPERATURE CORRECTION
      WILL BE COMPUTED FOR VARIABLE NUMBER AND KIND OF ATOMS, GIVING AT THE
      END THE RELIABILITY FACTOR. ALSO THE STRUCTURE FACTOR FOR DIFFERENCE FOU-
      RIER CAN BE COMPUTED. THE PROGRAM (WRITTEN IN ACODE) WITH DIRECTION OF
      USE IS PUBLISHED (ENGLISH) IN ACTA CHIM. HUNG., 1964.
6532 NE 803 B SPC      TOMAN, TICHY*   3, NORMAL BEAM                        LWA
6526 NE 803B   STAT    SASVARI         N(Z) FUNCTION FOR 2 AND 3 DIM DATA    LWA
      THE PROGRAM (WRITTEN IN ACODE) WITH DIRECTION OF USE IS PUBLISHED
      (ENGLISH) IN ACTA CHIM. HUNG., 1964.
6528 NE 803B   USF     SASVARI                                               LWA
      THE UNITARY STRUCTURE FACTORS WILL BE COMPUTED FROM THE ABSOLUTE ONES ON
      THE BASIS OF THE TABULATED UNITARY SCALE FACTOR VALUES FI. THE PROGRAM
      (WRITTEN IN ACODE) WITH DIRECTION OF USE IS PUBLISHED (ENGLISH) IN ACTA
      CHIM. HUNG., 1964.
7048 PALLAS    FR      PROTAS CNRS BEL PNAM ELECTRONIC DENSITY SECTIONS      MWA
7054 PALLAS    FR 2    RERAT CNRS BEL  P1                                    MWA
7049 PALLAS    ID      PROTAS CNRS BEL ID,BOND ANGLES,ANY SYSTEM             MWA
7037 PALLAS    LS 2 A  MARCHAN CNRSBEL PG1 REFIN.ATOM.COORD.ISOT.TR,SCL      MWA
7034 PALLAS    LS 2 C  FUERXER/LEROY   P2 REF.ATOM.COORD.ISOT.TF.SCL         MWA
7035 PALLAS    LS 2 C  PROTAS CNRS BEL P6 REFIN.ATOM.COORD.ISOT.TF.SCL       MWA
7036 PALLAS    LS 2 C  PROTAS CNRS BEL P6M REFIN.ATOM.COORD.ISOT.TF.SCL      MWA
7033 PALLAS    LS 2 C  SETI/CNRS       PGG PMG   REFIN. AT COORD,TF,SCL      MWA
7042 PALLAS    LS 3 A  CHANOINE BEL    B21M REFIN.AT.COORD.ISOT.TR.SCL R     MWA
7040 PALLAS    LS 3 A  PROTAS CNRS BEL P63 REF.ATOM.COORD.ISOT.TF,SCL        MWA
7039 PALLAS    LS 3 C  CHANOINE BEL    PNMA REF.AT.COORD.ISOT.TF.SCL         MWA
7055 PALLAS    LS 3 C  PROTAS CNRS BEL I41/ACD REF.ATOM.COORD.INDIV.TF.SCL   MWA
7041 PALLAS    LS 3 C  PROTAS CNRS BEL PNAM REFIN.AT.COORD.INDIV.ISO.TF.SCL  MWA
7038 PALLAS    LS 3 C  RERAT CNRS BEL  P21/NREF.AT.COORD.ISOT.TF.EA.AT.SCL   MWA
7056 PALLAS    PATSUP  RERAT CNRS BEL  PATT. SUPERPOS., USES FR P1 (7054)    MWA
7050 PALLAS    SF      FUER/LEROY CNRS SF,R, FOR P1 AND P1BAR GROUPS         MWA
7046 PALLAS    SPEC    CHANOINE BEL    P2/M ELECTRONIC DENSITY COEFFS.       MWA
7045 PALLAS    SPEC    CHANOINE BEL    PMMM PATT.COEF.10 SECT.               MWA
7051 PALLAS    SPEC    FUER/LEROY CNRS CALC ATOM SCATTERING FAC.FROM TABLE   MWA
7043 PALLAS    SPEC    FUER/LEROY CNRS CALC 1/D SQUARE,LF ALL SYSTEMS        MWA
```

```
7044 PALLAS     SPEC         FUER/LEROY CNRS     CALC 1/D SQUARE,SF,P1 AND P1BAR GRPS MWA
7053 PALLAS     SPEC         MARCHAN CNRS BEL    SCL AND TF WILSON METHOD            MWA
7052 PALLAS     SPEC         PROTAS CNRS BEL     NORMALIZE INTENS)FROM N FILMS NMAX20 MWA
7047 PALLAS     SPEC         PROTAS CNRS BEL     PNAM ELECTRONIC DENSITY COEFFS.     MWA
8524 PC1        D            IITAKA              1/D**2,2THETA(HKL) FOR ALL REFS     LWA
     INPUT, LC (OR RECIP. LC), WAVELENGTH, 2THETA MAX, EXTINCTION RULE.
     OUTPUT, LISTS H,K,L,Q,2THETA OR SINTHETA/LAM. FOR ALL REFS (EXCLUDING
     ABSENT REFS) WITHIN GIVEN 2THETA MAX.
8520 PC1        DP LP        IITAKA              CORR EQUI WEISS                     LWA
     LP CORRECTION FOR EQUIINC. WEISS. DATA. CORRECTS IOBS FOR ANY LAYER FOR A
     GIVEN AXIS (A OR B OR C).
8522 PC1        DP P         IITAKA              FIND NEAREST QCAL(HKL)FOR Q BY POWD. LWA
     STORE REC.LC, TABLE OF QOBS BY POWDER PHOTO, QOBS MAX. CALCULATES QCAL
     FOR ALL HKL WITHIN QMAX AND FINDS THE NEAREST QOBS. TABULATES HKL, QCAL,
     QOBS, QOBS-QCAL.
8525 PC1        ID           IITAKA              BOND SEARCHING. NO ANGLE CALTD.     LWA
     CALCULATES DISTANCES BETWEEN TWO ATOMS ONE IN AN ASYM. UNIT AND OTHERS IN
     THE SAME UNIT OR OUTSIDE OF THE UNIT. THE ATOMIC COORDINATES ARE DERIVED
     BY SYM. OPERATIONS INCLUDING LATTICE TRANSLATIONS. MAX AND MIN DISTANCES
     CAN BE DEFINED AND ALL THE DISTANCES WITHIN THE RANGE ARE PRINTED OUT
     TOGETHER WITH THE SYM. OPERATION CODES AND THE COORDINATES.
8521 PC1        SF           IITAKA              MONOCL.C.ISOTF FOR EACH ATOM KIND   LWA
     FOR MONOCLINIC CENTRIC SPACE GROUPS. ATOMIC SCATTERING FAC ARE CALTED BY
     THE 2 TERM GAUSSIAN EXPANSION. INPUT TAPE, HKL, SINTHETA/LAM, (FOBS),
     OUTPUT TAPE, HKL, SINTHETA/LAM, (FOBS), FCAL. MAX N=38-2J. SPEED=0.2N+
     0.4J SEC/PLANE. N ATOMS J KINDS IN AN ASYM. UNIT.
8523 PC1        TABLE        IITAKA              MAKE SINTHETA**2 OR Q TABL.VS.2THETA LWA
     MAKE SINTHETA**2(X10**5) OR QVALUE(X10**5) VS. 2THETA(X10**2) TABLES IN A
     CLEAR TABLE STYLE. STARTING AND ENDING 2THETA ANGLES CAN BE DEFINED. Q
     VALUES (1/D**2) ARE CALTED FOR ANY GIVEN WAVELENGTH.
3040 PDP 6      FR3          RAE                 IT=SFLS OT.OT FR/DIF                LPA
     A THREE PASS PROGRAM. FIRST PASS ACCEPTS OUTPUT FROM SLFS AND CONVERTS IT
     INTO A FORM SUITABLE FOR SECOND PASS WHICH EVALUATES A FOURIER OR DIFF.
     IN SECTIONS UP ANY AXIS AT INTERVALS MULTIPLES OF 120THS OF CELL EDGE.
     OUTPUT ON LINE LISTING AND TAPE. THIRD PASS ACCEPTS OUTPUT OF SECOND PASS
     AND CONTROUS FOURIER DIRECTLY ONTO A FAST CRT DISPLAY. PROGRAM
     LANGUAGE IS MIXED FORTRAN AND MACRO 6.
3039 PDP 6      SFLS         RAE                 TRICL,MONOCL,ORTH,AT                LPA
     PROGRAM IS IN TWO PASSES. 1ST PASS CALCULATES SFS AND LEAST SQUARES SUMS.
     SECOND PASS INVERTS LEAST SQUARES MATRIX AND EVALUATES NEW PARAMETERS.
     AND STANDARD DEVIATIONS. INDIVIDUAL ISOTROPIC OR ANISOTROPIC TEMPERATURE
     FACTORS MAY BE USED. ALLOWANCE MAY BE MADE FOR ANOMALOUS DISPERSION.
     THE BLOCK DIAGONAL APPROXIMATION IS USED. PROGRAM LANGUAGE IS FORTRAN.
3007 SILLIAC    ABS          LOVELL/FREEMAN      ABS CORR FOR XTLS WITH PLANE FACES  LWX
     ABSORPTION FACTORS FOR CRYSTALS BOUNDED BY PLANE FACES. AVAILABLE FOR
     THREE-DIMENSIONAL EQUI-INCLINATION DATA
3009 SILLIAC    ANGLES       LOVELL/FREEMAN      BOND ANGLES ONLY                    LWX
     COMPUTES ALL ANGLES SUBTENDED AT A CENTRAL ATOM BY PAIRS OF ATOMS FROM A
     SPECIFIED GROUP.
3008 SILLIAC    ID           LOVELL/FREEMAN      BOND DISTANCES ONLY                 LWX
     COMPUTES ALL DISTANCES LESS THAN A SPECIFIED LIMIT WITHIN BASIC CELL AND
     26 SURROUNDING CELLS. ALL SPACE GROUPS CAN BE HANDLED.
3010 SILLIAC    SF 3ATN      LOVELL/FREEMAN      ALL SYMM N/XDIFF DOES 2*2 FOR SCLTF LWX
     GENERAL STRUCTURE FACTPR PGM. AVAILABLE FOR ANY SYMMETRY. NEUTRON OR
     X-RAY DATA WITH ISOTROPIC OR ANISOTROPIC TEMPERATURE FACTORS. GIVES R
     FACTOR AND DOES 2X2 MATRIX FOR AVERAGE TEMPERATURE FACTOR AND SCALE
     FACTOR.
6506 URAL 1     DP           NOVAK               E101/LP,ABS.CORR                    LPA
6505 URAL 1     SF           LINEK,NVK,NADR      E82/GENERAL FORMULA.ICA             LPA
6507 URAL 1     STEREO       LINEK*NVK           FROM THREE FR2 A 3MODEL IS COMP.    NPX
     COMPUTATION OF THE CUBE ROOT OF PRODUCT OF RO(XY).RO(XZ).RO(YZ) IN EVERY
     (XYZ) FROM UNIT CELL.
8021 TR4-AGL    3 GEOM       SMITS               GRON-CH-9/MOLECULAR GEOMETRY        LPA
     BOND DISTANCES,BOND ANGLES,BEST PLANES,DISTANCES TO THESE PLANES
8022 TR4-AGL    LP ABS       VOS                 GRON-CH-10/ABS. CORR.               LPA
     LP AND ABS CORR.WEIS,PREC.+CAPILL,MAX.8X8X8 POINTS
```

```
8020 TR4AGL     PROF        BOOM              POLYCHR.D-SCH,KON.NED.AKAD.WET.68B46 LPA
     PROGRAM CALCULATES MONOCHROMATIC POINTFOCUS, POLYCHROMATIC POINTFOCUS,
     AND POLYCHROMATIC LINEFOCUS PROFILES FOR DEBYE-SCHERRER (PHOTOGRAPHIC)
     CYLINDRICAL POWDER CAMERAS FROM RADII AND ABSORPTIONCOEFFICIENTS OF SPEC
     IMEN AND MOUNTING CAPILLARY, PEAK WAVELENGTH, HALFWIDTH AND ASYMMETRY OF
     RADIATION USED, FOCUS WIDTH AND EMISSIVITY CURVE, AND CAMERA GEOMETRY,
     STORES PROFILE-COORDINATES ON MAGNETIC TAPE, PUNCHES INTERMEDIATE RESULTS
     IN PLOTTER-TAPE, INDICATES 'CORRECT' POSITIONS OF DIFFRACTION LINES. 23K.
6001 WGMATIC  LP WEIS3 ROMMING            CS-01/                                LWA
     LP OF INTENSITIES FROM WEIS ANY LAYER LINE. PRINTS SIN SQ THETA,
     CORRECTED INTENSITY, SQ ROOT OF CORRECTED INTENSITY.
8057 X1   8K       DP       RUTTEN-KEULEMAN OX4/DATA CONTROL                    MWA
8058 X1   8K       DP       RUTTEN-KEULEMAN OX5/AVERAGING DATA MAXES IT OX9     MWA
8055 X1   8K       FR 3     RUTTEN-KEULEMAN OX8/PATSUP TRICL MONOCL RHOMBIC     LWA
8056 X1   8K       SCL DP   RUTTEN-KEULEMAN OX3/LS-METH FROM DATA OF 2 AXES     MWA
8054 X1   8K       SFLS DM  RUTTEN-KEULEMAN OX9/AT ALL SPACEGR 3X3POS 6X6AT     LWA
     DIFFERENT SCALE FACTORS ALL KINDS OF SPECIAL POSITIONS CAN BE TREATED
     WITHOUT PROGRAMMING.
     GIVES INPUT FOR OX8 FOR RELIABLE REFLECTIONS.
3021 ZEBRA      DP          ROGERS         ZK3/CONVERT F COSALPHA TO FA,FB      MWA
     CONVERTS TAPES WITH /F/, COSALPHA TO FA, FB AND MARKERS FOR USE WITH
     ZK1/2.
3022 ZEBRA      DP          ROGERS         ZK5/OTHER OCTANTS                    LSW
     PREPARES 3D DATA FOR ZK1/2 FROM ABBREVIATED FORMAT. AVAILABLE FOR BOTH
     PMMM AND P212121.
3037 ZEBRA      DP SPC      ROGERS         ZK56/CENTROID DIFFR LINE             MNA
     EVALUATION OF SYMM FIXED-RANGE CENTROID OF DIFFRACTOMETER LINE (PIKE-
     WILSON METHOD).
3031 ZEBRA      ENERGY      ROGERS         ZK35/DIPOLE ENERGY IN TRICLINIC LATT MNA
     PROG FOR EVALTN OF DIPOLE INTERACTION ENERGY AS A FUNCTN OF RADIUS IN
     TRICLINIC LATTICE. USED FOR NYLON BUT GENERAL IN FORM.
8016 ZEBRA      FITPLANE PALM             0-519/FIT PLANE THROUGH SET OF POINTS LWA
3023 ZEBRA      FR OT       VAN DER SLUIS  ZK7/TRANSFORM FR OT                  LSW
     OFFERS 3 MODES OF DISPLAY OF FR SUMS TO FACILITATE CONTOURING.
8008 ZEBRA      FR 2 3      SMITS          ZK2/C TRICL. GRID 1/256              LWS
8007 ZEBRA      FR 2 3      SMITS,SCHOONE  ZK1/C OR A, TRICL. GRID 1/256        LWS
8012 ZEBRA      IT          SMITS          ZK23/PREPARES F-TAPES FOR ZK20,21,22 LWS
3012 ZEBRA,SC LIN REG       RUDOLPH          2 LINE LS, E                       LWA
     MAX 100 POINTS (WEIGHTS OPT), OT COEF LINEAR REGRESSION, VARIANCE
8015 ZEBRA      LP ABS.     PALM           0-500/F FROM I, EQUI-INCL WEIS       LWA
     CALC. F SQ, F, ABS.CORR. FOR SPHER. OR CYLINDR. SPECIMEN
8014 ZEBRA      LP WEIS     SMITS          ZK /F FROM I, ANTI-EQUI INCL         LWS
     CALCULATES SIN THETA, SIN SQ THETA, F SQ AND F FROM INTENSITIES
8013 ZEBRA      LP WEIS     SMITS          ZK43/F FROM I, EQUI-INCL WEIS        LWS
     CALCULATES SIN THETA, SIN SQ THETA, F SQ AND F FROM INTENSITIES
3018 ZEBRA,NC LS DATA       BOONSTRA       B05/PREPARES LS IT FROM PAT IT       NPA
     IT PAT TAPE ZK1,2. OT TAPE FOR IT LS ZK22. MANUAL CORRECTIONS NEEDED FOR
     CENTRAL LATTICE ROWS
8011 ZEBRA      LS          SCHOONE        ZK22/INDIV ISOTROPIC TF, DIAGONAL REF LWS
     INSERTS FOR VARIOUS SPACE GROUPS. MAX 64 ATOMS, 9 TYPES
3016 ZEBRA,NC MARGIN        BOONSTRA       B03/SHIFT TELEPRINTER MARGIN 10 SP   NPA
     SHIFT MARGIN OF TELEPRINTER OT 10 SPACES, EASILY CHANGED
3014 ZEBRA,SC P             PISTORIUS, C.  GI9/TAB P ACCORDING TO H             NPA
     AID TO PI. IT LC, OT OF P LISTED.ACC TO H. SUPERCEDED BY 704 PROG
3035 ZEBRA      PI          LOENE          ZK51/RECOG ZONES IN POWDER DATA      MPA
     USES DE WOLFFS METHOD.
3036 ZEBRA      PROF COR    NEETHLING      ZK54/STOKES CORRN FOR INST BROADENING LWS
3013 ZEBRA,SC PLANE         RUDOLPH          3 PLANE LS THRU ATOM POS           LWA
     MAX 100 POINTS (WEIGHTS OPT), OT LS PLANE, IND, TOTAL DEVIATION
     SUPERCEDED BY PROG MORGAN, QUEENS COLLEGE, DUNDEE
3017 ZEBRA,MC SF CORR       BOONSTRA,SMITS B04/SPOT-SHAPE INC IN WEIS INT CORR  NPA
     GRON-CH-20 BY SMITS (WEIS INT TO CORR SF) MODIFIED TO INCLUDE SPOT-SHAPE
     FACTOR
3015 ZEBRA,NC SF STAT       BOONSTRA       B02/SF, SF**2 INC SCL FROM LS ZK22   NPA
     DATA FOR STAT. IT FOR LS ZK22 USED TO LIST SF, SF**2 INC SCL
8009 ZEBRA      SF A        SMITS          ZK20/A INDIV ISOTROPIC TF            LWS
```

```
                INSERTS FOR VARIOUS SPACE GROUPS. MAX 80 ATOMS, 13 TYPES.
8010 ZEBRA      SF C    SMITS           ZK21/C INDIV ISOTROPIC TF              LWS
                INSERTS FOR VARIOUS SPACE GROUPS. MAX 80 ATOMS, 13 TYPES.
3033 ZEBRA      SPC     SMITS           ZK41/SPC FOR C2/C                      MNA
3019 ZEBRA      1FRTRANS NEETHLING      ZK70/1D FR TRANSFORMATION              MNA
                ACCEPTS 1000 ORDINATES, ADJUSTABLE SPECTRUM INTERVAL.
3027 ZEBRA      3 GEOM  ROGERS          ZK30/MOLECULAR BOND SCAN               LSW
                IDENTIFIES ALL INTERATOMIC DISTANCES WITHIN ASYMMETRIC LIMIT.  ALSO GIVES
                ORTHOGONALIZED COORDS IN AU.
3028 ZEBRA      3 GEOM  ROGERS          ZK31/INTERMOLECULAR CONTACTS SCAN      LSW
                IDENTIFIES ALL CONTACTS BETWEEN PARENT ASYMMETRIC UNIT AND NEIGHBORS WITH
                LENGTHS LESS THAN PRESCRIBED LIMIT.
3029 ZEBRA      3 GEOM  ROGERS          ZK32/MOLECULAR GEOMETRY                LSW
                GIVES BOND LENGTHS AND DIRN COSINES OF SPECIFIED BONDS, ANGLES BETWEEN
                STATED PAIRS OF BONDS, BEST PLANES (WITH OR W/OUT CONSTRAINTS),
                DISTANCES OF ATOMS FROM PLANES.
5034 Z22 R      ER2     BRUHN
5035 Z22 R      SF      TOEPFER,BRUHN
5071 Z23        FOLD    BRADACZEK,URBAN ELIM.LAV.SCATT.DET.ABSO.COEFF.CRYST.   LWA
                CALC. THEOR. CURVE  I = ALPH.I(L) + (1 - ALPH)I(B)
                I(B) = P(X) EXP(-UX) GAUSS    I(L) = I(B) 1/B.B
                PORT.LAV.SCAT. -- ANTL = ALPH.INT(I(L))/INT(I)
                PARAMETERS -- ALPH, U, G    G ELIM. PART. SIZE EFFECT
                MEAS. CURVE COMP. WITH I
                FOLD FOLDING INTEGRAL DETERMINATION
7568 Z23A 8K    FR 3 2  ZELENKO,ZAKRAJS GENL,60 GRID                           MPA
                FOR ZUSE Z23A, PAPER TAPE. ABS FRQ H,K,L LESS THAN 60. 2000 F(HKL), 8K.
                GENRALLY, ONE FR 3 DIMS WITH SYMMETRY ABOUT 3 DAYS.
5070 Z23        FT      LEMM            FT INTENS.LIQUIDS FOR ELIMIN.CUT-OFF   LPA
                CALCUL  1. R(RHO - RHO.ZERO) = F(S.ZERO)
                        2. R(RHO - RHO.ZERO) = F(R)
                        3. 4PI.R.SQU.RHO = F(R)
7567 Z23A 8K    LS      ZELENKO,LESJAK  GENL,WITHOUT WEIGHTING FUNCTION        MPA
                FOR ZUSE Z23A, PAPER TAPE. ABS FRQ H,K,L LESS THAN 60.
                2000 F(HKL), 8K. GENERALLY, FOR ONE ATOM IN 30 MINUTES.
7569 Z23A 8K    SF      ZELENKO,STALEC  GENL                                   MPA
                FOR ZUSE Z23A, PAPER TAPE. ABS FRQ H,K,L LESS THAN 60. 2000F(HKL), 8K.
                GENERALLY, ONE SF IN 5 HOURS
5069 Z23        STAT    LEMM            STRUCT.ANAL.DISTR.FUNCT.1-ATOM.LIQ.    LPA
                CALCUL. SYNTHETIC RAD. DISTRIB. FUNCTIONS OF LIQUIDS FROM CHOSEN ASYMM.
                DISTANCE STATISTICS FOR THE NEAREST NEIGHBOR BY CONVOLUTION-POLYNOM.
6518 ZRA 1      FR 2    WEISS           003*043                                LWA
                FR2 WITH DIFFERENT SUBPROGRAMS FOR PATTERSON,DIFFERENCE,FR WEIGHTED ACCOR
                DING TO ABSOLUTE VALUE OF SFOBS+SFCALC,AND OTHERS
6521 ZRA 1      ID      WEISS           001*022                                LWA
                ID OT VALUES SMALLER THAN LIMIT
6517 ZRA 1      LP,THETA KRAUSE         000*012                                LPA
                SINE THETA + LP FOR EQUIINCLINATION WEIS
6522 ZRA 1      SF,A    DENNER          002*121                                LWA
                SF+CONTRIBUTION OF SELECTED ATOMS/H/A
6515 ZRA1       SF,A,R  KRAUSE          002*023                                LWA
                SF + COMPARISON WITH SFOBS/R/A
6523 ZRA 1      SF,C    DENNER          002*111                                LWA
                SF+CONTRIBUTION OF SELECTED ATOMS/H/C
6524 ZRA 1      SF,C,R  KRAUSE          002*012                                LWA
                SF + COMPARISON WITH SFOBS/R/C
6516 ZRA1       SF,C,R  KRETSCHMER      002*211                                LPA
                SF/C/+COEFFICIENTS FOR DETERMINATION OF 3RD ATOM COORDINATE FROM KNOWN PR
                OJECTION AFTER KUTSCHABSKY/MONATSBER/ D/DAW/7/1965/95+ACTA/1965/IN PRESS
6519 ZRA 1      STAT    WEISS           000*306                                LWA
                STAT/HOWELLS ET AL ACTA 3,1950,210 FOR C-TEST
6520 ZRA 1      STAT    WEISS           000*309                                LWA
                STAT/SRINIVASAN ACTA 13,1960,388 FOR C-TEST


8053 1103       FR 3 2  KEUNING, VONK*  VOKAS/GRID N/10000                     LWA
```

```
8052 1103           FRSORTER  KEUNING, VONK*    VOPRE/PREPARES FOURIER DATA TAPE      LWA
8050 1103           LP        KEUNING, VONK*    VINTS/F FROM I                        LWA
7525 1103 2K        PAT2      MARIANI           FIAT INTP LANG PAT2 COS COS           LWA
8051 1103           SF        KEUNING, VONK*    VALCO/SF,SCL,TF,R                     LWA
7524 1103 2K        SF        MARIANI           FIAT INTP LANG SPACEGROUP 14 0 LEVEL  LWA
 493 1105           FR 2      MORROW            3/GRID N/50                           LWA
 494 1105           FR 3      MORROW            9/GRID N/360                          NPA
 495 1105           LP ABS    MORROW            14/LP OSC-ROT,WEISSENBERG,ABS 6PTGAUS NPA
 492 1105           SF        MORROW            2/GEN SF 100ATOMS ANISO TEMP FACTORS  LWA
 446 1604           FR 2 3    BLOUNT/DAHL*      FOURIER/FAST,GENERAL,BL,GRID N/240    LWA
```

446 1604 — APPLICABLE TO ANY SPACE GROUP THRU ORTHORHOMBIC (EXCEPT THOSE WITH D-GLIDE). USES UNIQUE REFLECTIONS ONLY. REGULAR OR DIFF FOURIER. VARIABLE OUTPUT FORMAT. PROGRAM IN CDC FORTRAN-63 AND CODAP-1 FOR USE WITH COOP MONITOR SYSTEM.

447 1604   FR3 SF   BLOUNT/DAHL*   PREFOUR/SF+FR WITHOUT SCRATCH TAPE   LWA
SAME AS FOURIER BUT ALSO INCLUDES SF AND LS SCALE FACTOR CALCULATIONS. UP TO 4000 INDEPENDENT REFLECTIONS STORED INTERNALLY. MAX 100 ATOMS IN ASYMMETRIC UNIT, OTHERS GENERATED BY SYMMETRY CARDS.

392 1604AFT2 SPEC   BROWN G M   ANALYZE/FIT OF F**2 CALC AND OBS   LPA
COMPUTES RATIO AVG F**2 OBS/CALC FOR RANGES OF F**2 OR OF INTENSITY EITHER CALC OR OBS OR FOR RANGES OF (SIN THETA/LAMBDA)**2. ALSO COMPUTES R FACTOR, AVG DELTA/SIGMA, ETC. INPUT TAPE FROM XFLS (AC 389) OR BMFLS (AC 390).

387 1604   STEREO   JOHNSON C K   ORTEP/THERMAL ELLIPSOID PLOT PROGRAM   LWA
UTILIZES PLOTTER SUCH AS CALCOMP TO PLOT BALL-AND-STICK CRYSTAL STRUCTURE FIGURES IN STEREO WITH THERMAL ELLIPSOIDS OR CIRCLES ON THE ATOMIC SITES. MAIN PROGRAM IS IN FORTRAN II. IBM 7090 OR CDC 1604 MACHINE ORIENTED PLOT ROUTINES ARE LIBRARY TYPE WITH MINOR MODIFICATIONS. 32K MACHINE REQUIRED.

523 1620T20K ABS   JOHNSON/PATT*   COMPUTES TRANSMISSION FACTORS   LWA
PROGRAM FOR CRYSTAL GROUND AS AN ELLIPSOID OF REVOLUTION. FOR G. E. GONIOSTAT TRANSMISSION FACTORS ARE COMPUTED AS A FUNCTION OF THETA AND CHI, FOR WEISSENBERG DATA AS A FUNCTION OF THETA AND LAYER LINE.

```
292 1620 SPS ABS BS    MAC/MOSELEY       ABS FOR GONIOSTAT.
294 1620 SPS AT CEN    MAC/LACHER        AT. CENTERS FROM 3-FR.
285 1620 40K D H       HAENDLER,COONEY   12/TETRAGONAL D SPACINGS              LWA
```
CALCULATES Q(HKL), SQUARE OF SIN THETA, D(HKL) FROM UNIT CELL DATA FOR HKL VALUES FROM 000 TO NNN, UP TO ANY GIVEN Q VALUE.

312 1620,704 D H LP   MEYER,MUELLER   D SPACE,HKL,PHI,CHI,LP,EXTINCTIONS   LPA
INPUT-TITLE CARD, CELL DATA, INSTRUCTION CARD. GENERATES INDICES WITH CHOICE OF EXTINCTIONS ACCORDING TO BUERGER (P83) CALCULATES D SPACINGS TOGETHER WITH TRIG. FUNCTIONS. PHI, CHI FOR 3D ORIENTATION,LP AVAILABLE

461 1620 20K D H PI   PIPPY/AHMED*   (HKL) + D(HKL) IN SPECIFIC SPHERE   LWA
CARD I/O, AUTOMATIC DIVIDE, INDIRECT ADDRESSES. GENERATES INDICES AND COMPUTES SPACINGS FOR ALL REFLEXIONS WITHIN ANY PART OF SPECIFIC SPHERE.

277 1620 40K D P   HAENDLER   2/ORTHORHOMBIC D SPACINGS   LWA
CALCULATES Q(HKL), SQUARE OF SIN THETA, D(HKL) FROM UNIT CELL DATA FOR HKL VALUES FROM 000 TO NNN, UP TO ANY GIVEN Q VALUE.

279 1620 40K D P   HAENDLER   4/MONOCLINIC D SPACINGS   LWA
CALCULATES Q(HKL), SQUARE OF SIN THETA, D(HKL) FROM RECIPROCAL CELL DATA FOR HKL VALUES FROM 000 TO NNN UP TO ANY GIVEN Q VALUE.

283 1620 40K D P   HAENDLER   9/FACE-CENTERED CUBIC D SPACINGS   LWA
CALCULATES Q(HKL), SQUARE OF SIN THETA, D(HKL) FROM UNIT CELL DATA FOR HKL VALUES FROM 000 TO NNN, WITH HKL ALL EVEN AND ALL ODD UP TO ANY GIVEN Q VALUE.

530 1620 40K D P   HAENDLER   8/SIMPLE CUBIC D SPACINGS   LWA
CALCULATES Q(HKL), SQUARE OF SIN THETA, D(HKL) FROM UNIT CELL DATA FOR HKL VALUES FROM 000 TO NNN UP TO ANY GIVEN Q VALUE.

531 1620 40K D P   HAENDLER   10/BODY-CENTERED CUBIC D SPACINGS   LWA
CALCULATES Q(HKL), SQUARE OF SIN THETA, D(HKL) FROM UNIT CELL DATA FOR HKL VALUES FROM 000 TO NNN, WITH H+K+L EVEN, UP TO ANY Q VALUE.

284 1620 40K D P   HAENDLER   11/HEXAGONAL D SPACINGS   LWA
CALCULATES Q(HKL), SQUARE OF SIN THETA, D(HKL) FROM UNIT CELL DATA FOR HKL VALUES FROM 000 TO NNN.

287 1620 40K D P   HAENDLER   14/TRICLINIC D SPACINGS   LWA
CALCULATES Q(HKL), SQUARE OF SIN THETA, D(HKL) FROM RECIPROCAL CELL DATA FOR HKL VALUES FROM 000 TO NNN UP TO ANY GIVEN Q VALUE.

```
456 1620 40K  DF 3       AHMED           GENERAL 3-D DIFFERENTIAL SYNTHESIS     LWA
     CARD I/O, AUTOMATIC DIVIDE, INDIRECT ADDRESSES. CALCULATES ELECTRON
     DENSITY, 3 FIRST AND 6 SECOND DERIVATIVES, FRACTIONAL SHIFTS AND PEAK
     CENTRES FOR UP TO 40 OR 50 ATOMS IN ONE PASS. ANY SPACE GROUP.
7545 1620 20K  DF 3      GIGLIO          GENERAL,FAST,AT MOST 32 ATOMS          LNA
7546 1620 20K  DF E      GIGLIO          STANDARD DEVIATIONS FROM DF            LNA
 454 1620 20K  DP WEIS   GABE/AHMED      INTENSITY ESTIMATE REDUCTION           LWA
     USES CARD I/O, AUTOMATIC DIVIDE, INDIRECT ADDRESSES. APPLIES SCL, LP,
     ABS TO WEISSENBERG MEASUREMENTS. FO SQ MAY BE SHARPENED.
 525 1620T20K  DP WEIS   MINKIN/PATT*    WEISSENBERG DATA REDUCTION             LWA
     CALCULATES STRUCTURE FACTORS FROM WEISSENBERG EQUI-INCLINATION DATA.
     CORRECTS FOR ABSORPTION BY INTERPOLATION FROM TABLE PRODUCED BY PGM 523.
 458 1620 20K  E         AHMED           COORDS S.D. SUMS + AGRMT ANALYSIS      LWA
     CARD I/O, AUTOMATIC DIVIDE, INDIRECT ADDRESSES. EVALUATES SUMS FOR S.D.
     OF ELECTRON DENSITY AND COORDS, R, SCALE. APPLIES TF CORRECTION. GIVES
     DETAILED ANALYSIS OF SF ERRORS. ANY SPACE GROUP.
 497 1620-2    FR        BRYDEN          2,3-DIM FOURIER SUM. CARD OUTPUT.      LPA
     COMPUTES 2 AND 3-DIM. FOURIER SERIES AT 120 DIV PER CELL EDGE, OR SUB-
     MULTIPLE OF 120.  CARD INPUT AND OUTPUT.
8045 1620C20K  FR        VDHELM,KING*    ERA210/FOURIER                         LWA
     CARD VERSION OF PROGRAMME 354
 485 1620 20K  FR2       HALL,SHIONO*    SPS,CARD, GRID N/100                   LWM
     TRICLINIC,MONOCLINIC,ORTHORHOMBIC, NO LIMIT ON 1ST INDEX,LESS THAN 31 FOR
     2ND. ANY NO. OF REFLEXIONS. IA,MF FEATURES.IBM LIBR. 8.4.005
 400 1620 SPS  FR2       SVETICH         012/112 GEN PROJ,N/200,PATT,F,DELTA F  LWA
     20K OR MORE, AUTO DIV. AVAILABLE W OR W/O INDIRECT ADD,PLEASE SPECIFY
     CARD I/O
 354 1620T20K  FR3       VDHELM/PATT*    GRID 1/100,OUTPUT NUM.OR ALPHAPLOT     LWA
     THE PROGRAM CALCULATES ONE SECTION AT A TIME. THE SORTING ORDER OF THE
     DATA ON THE INPUT TAPE DETERMINES ON WHICH AXIS THE SECTIONS ARE MADE.
     THERE IS A CHOICE OF A NUMERIC TYPOUT, DIRECT ALPHA NUMERIC TYPOUT ON THE
     APPROXIMATE SCALE OF THE UNIT CELL AND A POSSIBLE OUTPUT ON TAPE.
     THERE ARE NO SPACE GROUP LIMITATIONS.
 370 1620      FR 3 2    AHMED           GENERAL 3-D OR 2-D, GRID N/120         LWA
     GENERAL 3-D OR 2-D FOURIER. FOR 1620 WITH 40K OR 20K DIGITS, CARD I/O,
     AUTOMATIC DIVIDE+ INDIRECT ADDRESSES. INTERVALS OF N1,N2,N3/120. EMPLOYS
     THE DATA CARDS PRODUCED BY THE SF PROGRAM.
 486 1620 20K  FR3       HALL,SHIONO*    SPS,CARD, GRID N/100                   LWM
     TRICLINIC,MONOCLINIC,ORHTORHOMBIC, NO LIMIT ON 1ST INDEX,LESS THAN 31 FOR
     2ND. ANY NO. OF REFLEXION IN ANY ORDER. IA,MF FEATURES. IBM LIBR.8.4.006
     ALSO ALPHABETIC PLOTTING PROG. FOR CONTOURING, IBM LIBR. 8.4.007
 401 1620 SPS  FR PCH    SVETICH         220/225/240/250 FOURIER PCH OUT        LWA
     CC SS SC CS TABLE EXPANS AND PCH OUT FROM 1 QUAD CC SS SC CS TABLES IN
     STORAGE PCH 1,2 OR 4 QUAD OF THESE OR SUMS CC-SS OR SC+CS AND ADD ANY
     PREVIOUS PARTIAL CALC X1 OR X2 IN 20THS 25THS 40THS 50THS TO CONTOUR
     DIRECTLY ON 402 OR 407 PRINTOUT. AVAILABLE W OR W/O INDIRECT ADDR PL SPEC
 371 1620T20K  FRSORTER  VDHELM/PATT*    PREPARES FOURIER DATA TAPE             LWA
     THE PROGRAM PREPARES FOURIER AMPLITUDES FROM OUTPUT OF SFLS PROGRAM.
     IT REARRANGES THE TERMS IN EACH AMPLITUDE ACCORDING TO THE DESIRED
     SORTING ORDER. IT TAPESORTS THE AMPLITUDES.
8047 1620C20K  FRIT      KING            ERA235/CONVERSION ERA302 OT ERA210 IT  LWA
     TREATS SF OUTPUT FOR SPACE GROUP SYMMETRY.
 524 1620T20K  GONIO DP  JOHNSON/PATT*   G.E. GONIOSTAT DATA REDUCTION          LPA
     CONPUTES STRUCTURE FACTORS FROM DATA MEASURED WITH THE G.E. GONIOSTAT
     APPLIES ABSORPTION CORRECTIONS BY INTERPOLATION FROM TABLE PRODUCED BY
     PGM 523.  OUTPUT IS IN FORM SUITABLE FOR INPUT INTO PGM 372.
 280 1620 40K  H         HAENDLER,COONEY 5/INDICES TRANSFORMATION                LWA
     CONVERTS ANY SET (A) OF HKL VALUES TO SECOND SET (B) IF MATRIX OF
     TRANSFORMATION FOR SET A TO SET B IS GIVEN.
 281 1620 40K  H         HAENDLER,COONEY 6/REVERSE INDICES TRANSFORMATION        LWA
     CONVERTS ANY SET (B) OF HKL VALUES TO SECOND SET (A) IF MATRIX OF
     TRANSFORMATION FOR SET A TO SET B IS GIVEN.
8042 1620      H D       KING            ERA119/HKL + SIN ,HETA/LAMBDA          LPA
 499 1620-2    ID        BRYDEN          INTERATOMIC DISTANCES AND ANGLES       MPA
     COMPUTES INTERATOMIC DISTANCES AND ANGLES IN ALL CRYSTAL SYSTEMS.
     CARD INPUT AND OUTPUT.
```

```
 490 1620 20K  ID         CHU,SHIONO       SPS,CARD,ALL DISTANCES,ANY SYM.        LWM
     CALCULATE ALL DISTANCES LESS THAN A LIMIT. DIV,MF,TNF,IA FEATURES.
     ALL ANGLES AROUND AN ATOM BY 2ND PROG. IBM LIBR. 8.4.008
 526 1620T20K ID         JOHNSON/PATT*    INTERATOMIC DISTANCES AND ANGLES        LWA
     NO RESTRICTIONS ON LATTICE SYMMETRY.  UP TO EIGHT SYMMETRY-RELATED
     EQUIVALENT UNITS CAN BE SPECIFIED. ASYMMETRIC UNIT MAY CONTAIN ONE TO
     48 ATOMS.  LIMITS FOR DISTANCES AND ANGLES ARE SPECIFIED BY OPERATOR.
 459 1620 20K  ID (1)    PIPPY/AHMED*     BOND LENGTHS, ANGLES, E.S.D.S          LWA
     CARD I/O, AUTOMATIC DIVIDE, INDIRECT ADDRESSES. CALCULATES BONDS AND/OR
     ANGLES BETWEEN SPECIFIED ATOMS IN GIVEN LIST OF COORDS. CAN COMPUTE
     E.S.D.S FOR ANY ANGLE, AND ANY BOND EXCEPT IN TRICLINIC CELLS.
 460 1620      ID (2)    AHMED            INTER + INTRAMOLECULAR DISTANCES       LWA
     40K, OR 20K WITH MODIFICATIONS. CARD I/O, AUTOMATIC DIVIDE, INDIRECT
     ADDRESSES. FROM A LIST OF ATOMIC COORDS OF ONE MOLECULE, IT COMPUTES
     ALL BOND LENGTHS WITHIN A GIVEN LIMIT, ALL BOND ANGLES, GENERATES
     POSITIONS OF EQUIVALENT MOLECULES AND SCANS ALL INTERMOLECULAR DISTANCES.
 278 1620 40K  LC        HAENDLER,COONEY  3/DIRECT LATTICE FROM RECIPROCAL       LWA
     CALCULATES DIRECT LATTICE PARAMETERS FROM RECIPROCAL LATTICE PARAMETERS
     INCLUDES ANGLE FUNCTIONS AND VOLUMES.
 288 1620 40K  LC        HAENDLER,COONEY  16/RECIPROCAL LATTICE FROM DIRECT      LWA
     CALCULATES RECIPROCAL LATTICE PARAMETERS FROM DIRECT LATTICE PARAMETERS
     INCLUDES ANGLE FUNCTIONS AND VOLUMES.
 289 1620 40K  LC        HAENDLER,COONEY  19/AXIAL TRANSFORMATION                LWA
     CONVERTS 3-DIMENSIONAL COORDINATE SYSTEM (A) TO SYSTEM B, GIVEN THE
     MATRIX OF TRANSFORMATION FOR (A) TO (B). WILL CONVERT RECIPROCAL OR
     DIRECT CELLS.
 455 1620 40K  LF        AHMED            LF AND WILSON PLOT                     LWA
     CARD I/O, AUTOMATIC DIVIDE, INDIRECT ADDRESSES. COMPUTES SIN SQ THETA AND
     LF FOR MAX. OF 10 F-CURVES. EVALUATION OF TF AND SCALE BY WILSON PLOT
     OPTIONAL. PREPARES INPUT TO PROGRAM NO. 369.
8044 1620C20K LF         KING             ERA121/SUBTABULATION                   LWA
     INTERPOLATES PUBLISHED LF TABLES TO GIVE TABLES AT ANY DESIRED INTERVAL
     USES INDIRECT ADDRESSING
 488 1620 20K  LF        SHIONO           SPS,CARD,4 POINT INTERPLN.             LWM
     F VS. SIN OR SIN/LAMBDA TABLE OF DESIRED INTERVAL FROM LIT.VALUES.
     DIV,MF,TNF,IA FEATURES.  IBM LIBR. 8.4.003
 498 1620-2    LP        BRYDEN           CORR INT FROM EQUI-INCL WEISS          LPA
     CORRECTS EQUI-INCLIN WEISS PHOTOS FOR LP AND FOR UPPER-LAYER SPOT EXTEN
     BY THE METHOD OF PHILLIPS. CARD INPUT AND OUTPUT.
 489 1620 20K  LP        SHIONO           SPS,CARD,NORMAL-B./EQUI-INCLIN.        LWM
     WITH OR WITHOUT EXTENDED SPOT CORRECTION BY PHILLIPS.DIV,MF,IA FEATURES
     IBM LIBR. 8.4.002
 527 1620T20K LS OT IT   JOHNSON/PATT*    LS SUMS SOLVER - PARAMETER SHIFTER     LPA
     PARAMETER CORRECTIONS ARE COMPUTED BY SOLVING LEAST SQUARES SUMS WHICH
     ARE PART OF OUTPUT FROM PGM 372. NEW PARAMETER TAPE IS PREPARED FOR INPUT
     INTO ANOTHER CYCLE OF LEAST SQUARES WITH PGM 372.
 521 1620 40K  PI        HAENDLER,COONEY  POWDER INDEXING PROGRAM                LWA
     FROM ACTA CRYST., 16,1243(1963), BASED ON ITO METHOD
7547 1620 20K  R         DAMIANI          FAST,R FOR SYSTEMATIC TRANSLATION      MNA
     AND ROTATION IN THE SPACE OF ONE OR MORE ASYMMETRIC UNITS.THE NUMBER OF
     REFLEXIONS IS 120 AT MOST
3048 1620 60K  R         MASLEN           AGREEMENT ANALYSIS                     LWA
     R-FACTOR IS CALCULATED FOR SPECIFIED INCREMENTED RANGES OF FOBS,SIN
     THETA SQUD.,H,K,AND L. SLOWER FT2 VERSION FOR 20K MACHINE AVAILABLE
8048 1620C20K SCL TF     KING             ERA256/PATTERSON SHARPENER             LWA
     WILSON PLOT FOR SCL/TF. CALCULATES NORMALIZED STRUCTURE FACTORS AND
     COEFFICIENTS OF SHARPENED PATTERSON. USES INDIRECT ADDRESSING.
 402 1620 SPS  SF        CAUGHLAN         400/410 PARAMETER GEN AND SF CALC      LWA
     20K OR MORE, AUTO DIV.  CALC SF W ISO TEMP FACTOR. ANY SPACE GROUP
8043 1620      SF        KING             ERA164/TRICL                           PPA
 487 1620 20K  SF        SHIONO           SPS,CARD, TRI,MONO,ORTHO               LWM
     INDIV.ISOTROPIC. UP TO 150 ATOMS ONE PASS, ANY AXIAL SETTING, ANY NO.
     OF REFLEXIONS. DIV,IA,MF,TNF FEATURES. IBM LIBR. 8.4.004
 403 1620 SPS  SF        SVETICH          402/STRUCTURE FACTOR RESCALE           LWA
     20K OR MORE, AUTO DIV
7544 1620 20K  SF 3      DAMIANI          FAST,ISOTROPIC AND ANISOTROPIC TF      MNA
```

```
                NO GENERAL,AVAILABLES SPACE GROUPS 1,2,7,8,14 INTERNATIONAL TABLES
     8046 1620C20K SF LS      KING            ERA302/BLOCK DIAGONAL LS            LPA
                TRICLINIC,MONOCLINIC,ORTHORHOMBIC SPACE GROUPS. PARAMETERS = FSCALE,
                POSITIONS,ISOTROPIC TF. UP TO 68 ATOMS OF 9 TYPES IN ASYMMETRIC UNIT
                USES INDIRECT ADDRESSING.
      457 1620 40K SFLS       MAIR/AHMED      LS REFINEMENT. BLOCK DIAGONAL.      LWA
                CARD I/O, AUTOMATIC DIVIDE, INDIRECT ADDRESSES. COMPUTES SF AND
                ACCUMULATES LS TOTALS. GIVES NEW VALUES FOR ATOMIC COORDS, ISO AND/OR
                ANISOTROPIC  TEMPERATURE PARAMETERS, R, BONDS, ANGLES. ALL SPACE GROUPS
      369 1620     SF R SCL AHMED             SF+R+SCL ALL SPACE GRPS. VERY FAST  LWA
                SF R SCL. 1620 WITH 40K DIGITS, CARD I/O, AUTOMATIC DIVIDE + INDIRECT
                ADDRESSES. CALCULATES SF FOR ALL SPACE + PLANE GRPS. ALSO R + SCL.
                MAX. OF 50 ATOMS, 10 F-CURVES. ISOTROPIC T.F. AT PRESENT. VERY FAST.
      355 1620T20K SPC H   JOHNSON/PATT*   GONIOSTAT SETTINGS                     LWA
                G. E. GONIOSTAT ANGULAR SETTINGS FOR TRICLINIC OR HIGHER SYMMETRY.
                REQUIRES A REFLECTION AT CHI NINETY. OUTPUT SEQUENCE ARRANGED TO
                MINIMIZE DUPLICATE ANGLE SETTINGS.
      462 1620 20K SPC H LP PIPPY/AHMED*   (HKL) + GONIOSTAT SETTINGS + 1/LP      LWA
                CARD I/O, AUTOMATIC DIVIDE, INDIRECT ADDRESSES. GENERATES INDICES AND
                CALCULATES PHI, CHI, 2THETA, AND 1/LP FOR ALL REFLEXIONS WITHIN ANY
                PART OF SPECIFIC SPHERE IN RECIPROCAL SPACE. ALL SYSTEMS.
     8049 1620C20K SPC H      KING            ERA270/GONIOSTAT SETTER             LPA
                GONIOSTAT SETTINGS AND LP FOR ALL REFLEXIONS IN GIVEN SPHERE.
                USES INDIRECT ADDRESSING.
     5062 1620FTN2 SPEC       SEGMUELLER      SIMULTANEOUS BRAGG REFLEXIONS       LWA
                COMPUTES AZIMUTHS OF SIMULTANEOUS BRAGG REFLEXIONS (UMWEGANREGUNG)
                FOR ALL POSSIBLE PAIRS OF LATTICE PLANES FOR A GIVEN MAIN REFLEXION
                IN CUBIC AND HCP CRYSTALS. 40K.
     5011 2002      DIR       BEITINGER-HOP*  SIGN DET WITH DOUBLE PATT FUNCTIONS LXA
     5027 2002      FR 2,3    HILDEB,HAHN*    GEN VERS BL GRID VARIABLE           LWA
                FOR 2002 WITH PUNCHED CARD ACCESSORIES. SECTIONS PARALLEL
                A,B IN HEIGHTS Z. SIZE AND DENSITY OF GRID IN A,B AND
                Z-VALUES MAY BE SPECIFIED BY USER. ANY SYM. IT VIA CARDS,
                ONE FOR EACH H,K,L COMBINATION,SORTED ON L,K,H. OT ON DIRECT
                PRINTER. (SEE PGM NO. 5009)
     5053 2002      FR 2,3    PLEHWE,BAERNIG* BL. A OR C. FR 3 OR SECTIONS OR PRO- MWA
                JECTION OR PROJECTED SLABS PARALLEL X,Y. SIZE AND DENSITY OF GRID MAY
                BE SPECIFIED BY USER. BLK POSSIBLE. IT AND OT VIA TAPE.
                5000 WORD MAGNETIC CORE AND 10000 WORD DRUM MEMORY NECESSARY.
     5056 2002      H         PLEHWE,BAERNIG* H FOR THE 14 BRAVAIS LATTICES WITHIN MWA
                GIVEN LIMITING SPHERE, CALCULATION OF Q-VALUES FOR GIVEN RECIPROCAL
                LC AND SORTING BY INCREASING Q. OT HKL AND 4THETA OR SQARE OF
                SIN(THETA) OR 2SIN(THETA) OR D.
     5057 2002      ID        PLEHWE,BAERNIG* DISTANCES AND ANGLES BETWEEN        MWA
                SPECIFIED ATOMS ARE CALCULATED. ANY SYMMETRY.
     5054 2002      SF 3      PLEHWE,BAERNIG* A OR C OR INSERTS FOR SPACE GROUPS. MWA
                UNLIMITED NUMBER OF REFLEXIONS, UP TO 333 ATOMS AND 20 LF TABLES.
                INDIVIDUAL ISOTROPIC TF, IF NEEDED. R AND SCL, IF FO-VALUES ARE
                GIVEN. IT AND OT VIA TAPE.
     5055 2002      LP WEIS   PLEHWE,BAERNIG* LP CORRECTION FOR EQUI-INCLINATION  MWA
                WEIS DATA FOR ANY LAYER ABOUT ANY DIRECTION UVW. IT RECIPROCAL LC,
                ZETA UVW, HKL AND INTENSITY. OT SF AND SQUARE OF SF.
      267 205       EL        DARWIN, SMITH   TB71/RADIAL DISTR FUNCTION          LWM
      269 205       EL        DARWIN, SMITH   TB102/LEAST SQU REFINEMENT          LWM
      266 205       FR 2      EVANS/BURROUGHS TB59/PROJECTIONAL FOURIERS          LWM
                PROGRAM DEVELOPED BY US GEOLOGICAL SURVEY AND DISTR BY BURROUGHS CORP.
                OUTPUT IS ON 407 TABULATOR.
      114 205       FR 3      MARSH,EICHHORN  TB22/FOURIER AND VECTOR MAPS        LWM
                GENERAL FOURIER SYNTHESIS ACCORDING TO BEEVERS-LIPSON PRINCIPLE,
                IN UP TO 76-TH CELL EDGE FOR ALL SPACE GROUPS. PRINTS FIGURE FIELD
                DIRECTLY, NEGATIVE VALUES IN RED. PERMITS FORMATING FOR OBLIQUE CELLS
                AND APPROXIMATE RATIO OF EDGES.
      111 205       H         EICHHORN        TB19/LATTICE POINT GENERATION       LWM
                PROGRAM GENERATES ALL POSSIBLE INDEX COMBINATIONS WITHIN GIVEN LIMITING
                SPHERE, SUBJECT TO SPACEGROUP CONSTRAINTS. PRINTS LIST OF PERMITTED
                (HKL) WITH THEIR SINESQUTHETA VALUES.
```

```
112 205      ID E    EICHHORN         TB20/BONDS,ANGLES AND ERRORS         LWM
    GIVEN ATOMIC POSITIONS WITH THEIR ESD VALUES, COMPUTES ALL BONDS THAT
    ARE LESS OR EQUAL TO SUPPLIED MAXIMUM VALUE. CALCULATES MOST PROBABLE
    ERROR OF DISTANCES. THEN COMPUTES REQUIRED ANGLES AND THEIR MPE.
113 205      LP 3    EICHHORN         TB21/LP FOR PRECESSION               LWM
    PROGRAM COMPUTES LP FACTORS ACCORDING TO WASER FORMULA, THEN PROCESSES
    INTENSITIES AND PRINTS LIST OF (HKL),RAW INT, FSQU, F AND SINESQUTHETA.
115 205      LP 3    EICHHORN         TB23/LP OSCILLATION-ROTATION-WEISSBG LWM
    PROGRAM USES DIRECT EXPRESSION FOR LP FACTOR, HAS OPTION FOR ROTATION-
    OSCILLATION OR WEISSENBERG EQUI-INCLINATION CASE. PROCESSES RAW INT DATA
    AND FURNISHES LIST OF (HKL), INT, F, FSQU AND SINESQUTHETA.
260 220      DF      EICHHORN         TB58A/P-BAR-ONE DIFF SYNTH           LWM
    COMPLETE MATRIX WITH INDIVIDUAL ANISOTROPIC DEBIJE-WALLER FACTORS.
    PROGRAM CAN BE OVERLAID FOR OTHER CENTRIC SPACEGROUPS. DEPENDING UPON
    SYSTEM MAKEUP CAN ACCOMODATE UP TO 99 ATOMS OF UP TO 8 SCATTERING TYPES.
    NO RESTRICTION ON NUMBER OF REFLEXIONS.
    PROGRAMS TB 58B THROUGH G DOVETAIL INTO TB 58A.
298 220      DP      SILVERTON        X-2A/LP + PREP DATA TAPE FOR =117    LPA
265 220      FR 3    HOOGSTEEN,EICHH  TB58G/FOURIER AND VECTOR MAPS        LWM
    PROGRAM CONTAINS SCHOMAKER-SHARPENING SUBROUTINE FOR PATTERSON VECTOR
    MAPS. OUTPUT IS ON 407 TABULATOR. GRID INTERVALS UP TO 1/100 OF EDGE.
300 220      FR 3    SILVERTON        X-4C/ADAPTATION PROG 2 TO 3          LPA
    PRELIM. 1D SUMMN. FOR ANY MONOCLINIC CENTRIC SP GP.
    TO ADAPT HOOGSTEEN 2D PROG. FOR 3D
264 220      H S     EICHHORN         TB58F/UNIT CELL AND LATTICE POINT GEN LWM
    FIRST COMPUTES RECIPROCAL CELL FROM DIRECT CELL, OR VICE-VERSA.
    THEN GENERATES, SUBJECT TO SPACEGROUP CONSTRAINTS, ALL POSSIBLE (HKL)
    WITHIN GIVEN LIMITING SPHERE.
261 220      ID E    EICHHORN         TB58B/POSITIONS,BONDS,ANGLES AND ESD LWM
    PROGRAM FIRST COMPUTES ESD FOR ATOMIC POSITIONS ACCORDING TO CRUICKSHANK
    EXPRESSION, THEN COMPUTES DISTANCES AND ANGLES AND THEIR MPE.
262 220      LF      EICHHORN         TB58C/PREP OF SCATT CURVES           LWM
    CURVEFITS GENERALLY SUPPLIED (INT.TABLES,ETC) LITTLE F TO YIELD A SET OF
    PUNCHED CARDS AS STANDARD INPUT FOR TB 58A.
116 220      LF S    EICHHORN         TB41/HARTREE WAVEFIELD TO LITTLE F   LWM
    INPUT TO PROGRAM ARE SELFCONSISTENT FIELD WAVEFUNCTIONS WITH P AS
    FUNCTION OF R IN BOHR RADII. ROUTINE INTEGRATES WAVEFUNCTIONS PER SHELL,
    MULTIPLIES INDIVIDUAL VALUES BY ELECTRON OCCUPANCY, TOTALS FOR ALL SHELLS
    AND PRINTS ATOMIC SCATTERING FUNCTION AT INTERVALS OF 0.02 IN SINETHETA
    OVER LAMBDA. INTEGRATION PROCEDURE IS BY OVERLAP-CURVEFITTING.
263 220      LP 3    EICHHORN         TB58E/LP OSCILLATION-ROTATION-WEISSBG LWM
    PROGRAM USES DIRECT EXPRESSION FOR LP FACTOR, HAS OPTION FOR ROTATION-
    OSCILLATION OR WEISSENBERG EQUI-INCLINATION CASE. PROCESSES RAW INT DATA
    AND FURNISHES LIST OF (HKL), INT, F, FSQU AND SINESQUTHETA.
268 220      LS S    HEBERT/BURR      TB97/LS PLANE WASER-MARSH METHOD     LWM
    PROGRAM WILL ACCEPT A MAXIMUM OF 50 POINTS, WILL FIRST ORTHOGONALIZE
    POSITIONS AND THEN COMPUTE OPTIMUM LEAST SQUARES PLANE.
297 220      SPC     SILVERTON        X-1/GONIOSTAT SETTINGS               LPA
7563 6001 1K AT      MUSATTI          SHIFTS OF AT                         LPA
    CALCULATES THE SHIFTS OF ANISOTROPIC THERMAL PARAMETERS FROM OBS AND CALC
    DERIVATIVES BY NARDELLI-FAVA METHOD. THIS METHOD MAKES CALCULATION
    INDEPENDENT OF THE NUMBER OF REFLECTIONS.
7565 6001 1K AT SD   ANDREETTI        ANISOTROPIC THERMAL PARAM SD         LPA
    CALCULATES SD OF ANISOTROPIC THERMAL PARAMETERS BY CRUICKSHANK FORMULAE.
7561 6001 1K DF 2 3  MUSATTI          GEN 2 AND 3D DF                      LPA
    CALCULATES THE ELECTRON DENSITY, 3 FIRST DERIVATIVES AND 6 SECOND
    DERIVATIVES AT ASSUMED ATOM POSITIONS (NO RESTRICTIONS ON NUMBER OF
    REFLECTIONS AND OF POSITIONS) AND THE ATOMIC SHIFTS FOR OBS AND CALC
    DATA IN ORTHORHOMBIC, MONOCLINIC, AND TRICLINIC SPACE GROUPS.
7566 6001 15 DP      ANDREETTI        LS PLANE POL COORD BONDS ANGLES      LPA
    CALCULATES WEIGHTED PLANE THROUGH GIVEN POINT USING LS METHOD, AND
    DETERMINES STATISTICAL SIGNIFICANCE OF DISTANCES FROM THIS PLANE. CAL-
    CULATES POLAR COORDS. FOR REPRESENTATION OF ATOMIC ENVIRONMENTS.
    GIVEN ATOMIC POSITIONS WITH THEIR E.S.D. VALUES, COMPUTES ALL BOND
    DISTANCES THAT ARE LESS THAN OR EQUAL TO SUPPLIED MAXIMUM VALUE, AND
    ALL POSSIBLE ANGLES. E.S.D. OF BOND DISTANCES AND ANGLES ARE CALC.
```

```
                    WITH DARLOW FORMULAE.
7506 6001  1K  DP          PANATT/LOMBARD*   RESCALING FOR SF                  LWM
7509 6001  1K  DP PREC     PANATTONI,ZANO*   LP LF THETA                       LWA
7512 6001  3K  DP SCLTF    PANATTONI,ZANO*   WILSON PLOT RESCALING             LWA
7507 6001  1K  DP WEIS     PANATTONI,ZANO*   LP LF THETA                       LWA
7502 6001  1K  FR 2 A      PANATT/DEGRIF*    FR BLK                            LWM
7501 6001  1K  FR 2 C      PANATT/DEGRIF*    FR BLK                            LWM
7503 6001  1K  FR 3        PANATT/LOMBARD*   FR BL CENTRIC/ACENTRIC            LWM
7554 6001  1K  FPK         DOMENICI,ZUCCA    FPK AND STOKES CORRECTION         LWA
           CONVERTS THE 2 THETA DATA IN FUNCTION OF S, PERFORMS FOURIER INVERSION OF
           THE EXPERIMENTAL PEAK SHAPE AND OF THE STANDARD PEAK, GIVES AVERAGE
           CRYSTALLITE SIZE AND THE CORRECTED PEAK SHAPE BY FURTHER INVERSION.
7553 6001  1K  FT          DONENICI,ZUCCA    FT FUNCT.R,S MAX,CORR.TERM.EFFECTS LWA
           CONVERTS THE 2 THETA DATA IN FUNCTION OF S, 2 WAVE LENGTHS ACCEPTED,
           PERFORMS FOURIER INTEGRAL AS FUNCTION OF R AND/OR S MAX, CORRECTS
           TERMINATION EFFECTS BY EXTRAPOLATION.
7555 6001  1K  ID          DOMENICI,MIGLIO   RDF,PERFECT CRYSTAL,MULTI ATOMS    NPA
           GIVES A LIST OF THE INTERATOMIC DISTANCES FROM R = 0 TO R = 15 ANGSTROMS,
           ARRANGED ACCORDING TO THE NUMBER AND TYPE OF COORDINATED ATOMS. PRESENTLY
           ABBREVIATED FOR SIMPLE CASES.
7558 6001      .LP         DOMIANO           SPOT EXTEN. LP CORR 3 AXES         LPA
           CALCULATES LORENTZ POLARIZATION AND REFLECTION SPOT EXTENSION CORRECTIONS
           FOR REFLECTIONS TAKEN ABOUT 3 AXES.
7556 6001      LS C        PANATT/GRUB       LS/3D BLOCK 3X3 6X6 1X1 AT         LWA
           ALL SPACE GROUPS, NO LIMIT ON ATOMS AND REFLECTIONS, PARAMETERS,POSITIONS
7564 6001  1K  PARAM SD MUSATTI              POS PARAM SD                       LPA
           PROGRAM CALCULATES STANDARD DEVIATIONS OF ELECTRONIC DENSITY AND OF FIRST
           AND SECOND DERIVATIVES AND OF POSITIONAL PARAMETERS ON A GIVEN ATOMIC
           PEAK.
7559 6001      SCL LS      DOMIANO           LS SC FAC DETN FOR MAX18LAYERS 2AX LPA
           CALCULATES INTERLAYER SCALING FACTOR FOR TWO AXES DATA BY LEAST SQUARES
           FROM EIGENVALUES AND EIGENVECTORS OF SYMMETRIC MATRIX UP TO ORDER 18.
           CALCULATES SCALED AND AVERAGED INTENSITIES FOR REFLECTIONS OBS ON MORE
           THAN ONE LAYER.
7511 6001  3K  SCLTF LS    PANATTONI,ZANO*   RESCALING SF                       LWA
7504 6001  1K  SF 3 A      PANATT/DEGRIF*                                       LWM
7505 6001  1K  SF 3 C      PANATT/DEGRIF*                                       LWM
7562 6001  1K  SFRS R      MUSATTI           SF RESC AND R DETN                 LPA
           RESCALES SF AND CALCULATES R FOR ALL REFLECTIONS OR FOR SINGLE LAYERS.
7560 6001  1K  TF LS       ANDREETTI         SCL AND TF BY LS FROM WLP          LPA
           SCALE AND TEMPERATURE FACTORS ARE CALCULATED FROM THE LEAST-SQUARES
           STRAIGHT LINE.
  34  650      ABS         SANDS             ABCOR/PLOTS ABS COR TEMPLATE FOR WEIS
5029  650      DP          BARTL             SCL AND OVERALL TF (WILSON'TEST)   LWA
           PREPARES DATA FOR WILSON'PLOT. K AND B MUST BE EVALUATED GRAPHICALLY.
           PROGRAM THEN APPLIES SCL AND TF TO LP'CORRECTED INTENSITIES.
5028  650      DP          BARTL             PL AND ABS CORRECT FOR EQUI WEIS   LWA
           CORRECTS I OBS FOR LP'FACT. ABSORPTION
           CORRECTION BY MEANS OF TABLE OF A(SIN THETA)
           SUPPLIED BY USER COMPUTES ALSO SIN THETA.
           APPLICABLE FOR EQUIINCLINATION WEIS. (MONOCLINIC AND HIGHER SYM).
 185  650      DP          BROWN/LINGFTR*/5306 WILSON PLOT, SCL AND TEMP FACTOR MWA
           LIMITED TO MONOCLINIC OR TRICLINIC SYMMETRY
 180  650      DP          BROWN,STEWART/LINGFTR*/530X DATA REDUCTION           LWA
           5301,5302/FORM FACTOR INTERPOLATION AND STORAGE FOR DATA REDUCTION
           5303/DATA REDUCTION, LP 3 WEIS, MONOCLINIC SYMMETRY AND HIGHER
           5304/DATA REDUCTION, LP 3 WEIS, ALL SYMMETRIES
           5305/DATA REDUCTION, LP 2 PREC, MONOCLINIC SYMMETRY AND HIGHER
5036  650      DP          DIETRICH          X1/SHARPENING (VARIABLE FUNCTION)  MWA
           IT  CONSTANTS FOR SHARPENING FUNCTION. OT OF (5040) OR (5041)
           OT  (5040) OR (5041) CARDS RESPECTIVELY, CONTAINING SHARPENED FO
5041  650      DP          DIETRICH          X10/INTERPOLAT SCAT FAC,CORREL LAYERS MWA
           IT  SCL FOR LAYER, UP TO 10 FORM FACTOR TABLES FROM LITERATURE (WIDE
               INTERVALS, SIN THETA/LAMBDA, F), OT OF (5040) COMPLETED BY PUNCHING
               ABS FACTOR AND WEIGHT, AND/OR OT OF (5039)'
           OT  FOR (5040)'CARDS   INDICES, SIN**2 THETA/LAMBDA**2, WEIGHT, FO, UP TO
```

```
              10 SCATTERING FACTORS (ACCURATE LOGARITHMIC INTERPOLATION)
              FOR (5039)'CARDS   SAME AS ABOVE EXCEPT WEIGHT AND FO
5042 650      DP      DIETRICH        X11/SELECTS DATA, SCALES, 1ST SUMM FR MWA
      IT  1 CARD WITH 30 STATEMENTS ABOUT WANTED TYPE OF SYNTHESIS, GRIDS AND
          LOWER AND UPPER LIMITS FOR X,Y,Z, SUMMATION ORDER, CONDITIONS FOR THE
          SELECTION AND PROCESSING OF DATA. 1 CARD WITH 21 FURTHER STATEMENTS
          (CHANGABLE DURING RUN) ABOUT CONDITIONS FOR SELECTION AND PROCESSG OF
          DATA. OT OF (5037) OR (5038) OR/AND PREVIOUS OT OF (5042)FOR ADDITION
      OT  COMPLETE AND OPTIMAL IT FOR FR'PROGR (5047) FOR SECTION OR PROJECTION
          EXTRA CARD CONTAINING CHECK TOTALS, LARGEST COEFFICIENT
5044 650      DP      DIETRICH        X13/N(Z)'TEST                      MWA
      IT  LIMITS OF RECIPROCAL SPACE OF DATA, 2 STATEMENTS FOR AUTOMATIC
          SELECTION OF (OVERLAPPING) RANGES OF SIN**2 THETA/LAMBDA**2, SCL,
          OT OF (5040) OR (5041)
      OT  N(0.1)TO N(1.0)AND MEAN VALUES OF FO**2 AND SIN**2 THETA/LAMBDA**2
          FOR ALL RANGES OF THE LATTER. MEAN VALUES OF N(Z) AND DIFFERENCES TO
          THEORETICAL VALUES FOR ACENTRIC AND FOR CENTERED STRUCTURE
5046 650      DP      DIETRICH        X15/SELECTS DATA,PRODUCE IT FOR( 118) MWA
      IT  8 STATEMENTS FOR SELECTION OF DATA AND TYPE OF SYNTHESIS WANTED,F000,
          VOLUME OF CELL, 2 SCL ONE OF WHICH CAN BE CHANGED DURING RUN, OT OF
          (5040), (5037) OR (5038)
      OT  COMPLETE IT FOR ( 118)
          2 EXTRA CARDS CONTAINING MAXIMUM AND MINIMUM VALUES OF OT
 181 650      DP ABS  STEWART/LINGFTR*/5308 ABSORPTION CORRECTION        MPA
          LIMITED TO ZERO LEVEL WEIS DATA, RECTANGULAR CRYSTAL CROSS SECTION
5040 650      DP LP   DIETRICH        X9/CORRECTS EQUIINCL WEIS INT,SQ ROOT MWA
      IT  WAVE LENGTH, EQUIINCLINATION ANGLE, OT OF (5039), COMPLETED BY
          PUNCHING INTENSITIES
      OT  INDICES, SIN**2 THETA/LAMBDA**2, FO, FO**2, 1/LP, SQUARE ROOT OF LP
 324 650      DP LP   SHIONO          INTENSITY CORR. WEISSENBERG        MWA
          BASIC 650. EQUI-INCL WEISS. PRINCIPAL OR NON-PRINCIPAL AXES ROTN
 329 650      DP PATS SHIONO          SHARPENED PATTERSON COEF PGM170,171 LWA
          BASIC 650
5038 650      DP R    DIETRICH        X6/SCALES FO. AO,BO,FO-FC,AO-AC,BO-BC MWA
      IT  OT OF (5037)
      OT  INDICES, FO,FC, FO-FC. FOR SYMMETRY P1 ALSO AO, BO, AO-AC, BO-BC
          3 ADDITIONAL CARDS CONTAINING 2 TYPES OF R AND MAXIMUM AND MINIMUM
          VALUES OF MAIN OT
 325 650      E       SHIONO          STAND.DEV OF ELECTRON DENSITY ETC   LWA
          BASIC 650
 302 650      FPK LP  DEANGELIS/COHEN BL/FOURIER ANAL.OF PEAK SHAPES,STOKES LWA
          COMPUTE OR READ IN SIN AND COS COEFF. OF ANNEALED PEAK. COMPUTES SIN AND
          COS STOKES CORR. COEFF. OF BROADENED PEAK AND VALUES OF SIN/COS. CORRECTS
          FOR VARIATIONS OF  ANGULAR FACTORS ACROSS PEAK. OUTPUT 31 COEFF. AT 1/2
          HARMONIC INTERVALS AND 9 COEFF. AT 1/4 INTERVALS FOR ALL PEAKS.
 508 650      FR 2    BAUR            GENL PROJ CC SS CS SC               LWA
          650 WITH INDEX REG, VARIABLE GRID.
 356 650      FR PCH  JENSEN          CC SS SC CS TABLE EXPANS AND PCH OUT LWA
          FROM 1 QUAD CC SS SC CS TABLES ON DRUM PCH 1 2 OR 4 QUAD OF THESE OR SUMS
          CC-SS OR SC + CS AND ADD ANY PREVIOUS PARTIAL CALC X2 OR X3 IN 20THS
          25THS 40THS 50THS CONVENIENT TO CONTOUR DIRECTLY ON 402 OR 407 PRINT OUT
5047 650      FR 2    DIETRICH        X22/GENERAL,FAST, WITH( ) SECTIONS MWA
      IT  OT OF (5042) AND EVENTUALLY OT OF PREVIOUS RUN OF (5047) FOR ADDITION
      OT  1 HEADING' AND CHECK'CARD. SYNTHESIS (7 VALUES PER CARD)
 170 650      FR 2    SHIONO,HELLNER,WOLFEL/GENL PROJ CC SS CS SC, N/250  MWA
          FOURIER PROJECTION ALSO FOR PGM 171. FOR BASIC 650 OR WITH IXA, IAS AND
          UP TO 4 TAPES OR IXA, IAS, TAPES AND ON-LINE 407
 172 650      FR2     JENSEN          ENL PROJ,N/200,PATT,F,DELTA F       LWA
 171 650      FR 3    SHIONO          FOURIER,PATT,DIFFERENCE N/60        MWA
          MONOCL,ORTHO. USE OUTPUT OF PGM 176. PGM 170 USED FOR 2ND,ORD SUMS.
          IXA, IAS, UP TO 5 TAPES, ON-LINE 407(OPTIONAL).
 198 650      FR3     JENSEN          GENL,N/1000XN/200,PATT,F,DELTA F    LWA
 174 650      FR 3 DF SHIONO          DIFFERENTIAL SYNTH                  LWM
          MONOCL,ORTHO. USE OUTPUT OF PGM 176. BASIC 650 OR WITH IXA,IAS, 1 TAPE.
5039 650      H       DIETRICH        X8/LATTICE POINT GENERATION         MWA
      IT  A*,B*,C*,ALPHA*,BETA*,GAMMA*,MAXIMUM OF SIN THETA/LAMBDA,TYPE OF DATA
```

```
                    WANTED (1,2,3 DIMENSIONAL, INDICES,FOR WHICH ALSO NEGATIVE VALUES
                    HAVE TO BE CONSIDERED)
             OT   INDICES, SIN**2 THETA/LAMBDA**2. NO POINTS RELATED BY FRIEDEL LAW
  290 650         H        SVETICH          2010/LIMITING SPHERE REFLNS T,M,O      LWA
             PROGRAM CALCULATES AND PUNCHES ALL INDEPENDENT REFLECTIONS IN LIMITING
             SPHERE OF REFLECTION FOR TRICLINIC MONOCLINIC AND ORTHORHOMBIC SYSTEMS
             SECOND SETTING ONLY FOR MONOCLINIC SYSTEMS
  183 650         HYDROGEN STEWART,LINGFTR*/5105 H COORDS,BOND LENGTHS,ANGLES      LWA
             CALCULATES BOND LENGTHS AND ANGLES, GENERATES HYDROGEN POSITIONS FOR
             TERMINAL AND CHAIN TETRAHEDRAL, TRIGONAL, TERMINAL, OR H-BONDED
             ATOMS FOR SPECIFIED ATOM-HYDROGEN DISTANCES, MONOCLINIC OR HIGHER
  221 650         ID       LINGFTR*,BROWN/5112 INTERMOLECULAR DISTANCES            LWA
             GENERATES AND STORES RELATED ASYMMETRICAL UNITS SPECIFIED AND CALCULATES
             ALL INTER- AND INTRA-ATOMIC DISTANCES WITHIN MIN-MAX SPECIFIED LIMITS
  184 650         ID       SHIONO             DIST,ANGLES,S.D. OF ANGLE, ANY SYM   LWA
             BASIC 650
  220 650         ID       STEWART,LINGFTR*,BROWN/5102 BOND LENGTHS, ANGLES        LWA
             CALCULATES BOND LENGTHS AND ANGLES BETWEEN SPECIFIED ATOMS WHOSE
             COORDINATES ARE STORED IN MEMBORY, MONOCLINIC OR HIGHER SYMMETRY
  327 650         LF       SHIONO              4 POINT INTERPOLATION, PGM 176      LWA
             650 WITH IXA., ON-LINE 407(OPTIONAL)  PREPARES ATOMIC F TABLE
  133 650         PABSPROF MUSILBEUTHOMAS  GOODYEARATOMICREPORTGATDM829DECK806     LWA
             PROGRAM WRITTEN FOR IBM 650 TO CALCULATE COMBINED ABSORPTION AND
             RADIAL DIVERGENCE CORRECTIONS FOR DEBYE-SCHERRER POWDER DIFFRACTION
             LINES.  PROCEDURE THAT OF TAYLOR AND SINCLAIR )PROC. PHYS. SOC.
             LONDON,57, 1945) FOR DIVERGENT RADIATION.  MACHINE CALCULATION TIME
             ABOUT TWO AND ONE-HALF HOURS.  INPUT DATA. SAMPLE RADIUS, TARGET TO
             SAMPLE DISTANCE, CAMERA RADIUS, MEASURED 2Q, LINEAR ABSORPTION
             COEFFICIENT OF SAMPLE.  PROGRAM OUTPUT IS RELATIVE INTENSITY VS. 2Q,
             REFERENCED TO TRUE 2Q POSITION.
  509 650         LP       BAUR                PREC ANY LEVEL                      LPA
             650 WITH INDEX REG AND FLOAT,PT., ANY SYMMETRY.
  182 650         PLANE    STEWART/LINGFTR*/5103 LEAST SQUARES PLANE               LWA
             CALCULATES LS PLANE FOR MONOCLINIC AND HIGHER SYMMETRY, RESTRICTED TO
             TOTAL OF 98 ATOMS AND 0.0035A MIN PLANE-ORIGIN DISTANCE
  218 650         R SCL    BROWN,STEWART/LINGFTR*/440X FO DATA RESCALING           LWA
             4400,4401/FO RESCALING AND STATISTICS PROGRAMS
             4402/APPLIES OVERALL ISOTROPIC TEMP FACTOR CORRECTION BEFORE RESCALING
 5043 650         SCL TF   DIETRICH          X12/WILSON PLOT                       MWA
             IT   LIMITS OF RECIPROC. SPACE OCCUPIED BY DATA,2 STATEMENTS FOR AUTOMATIC
                  SELECTION OF AND OVERLAPPING OF INTERVALS OF SIN**2 THETA/LAMBDA**2,
                  OT OF (5041)
             OT   ALL POINTS OF WILSON PLOT, SCL AND TF CALCULATED BY FITTING STRAIGHT
                  LINE TO THE POINTS BY LS
 5037 650         SF AT    DIETRICH          X5/CALCULATES ALSO SCL                MWA
             IT   COORDINATES OF UP TO 200 ATOMS, CONSTANTS FOR UP TO 58 AT AND FOR UP
                  TO 42 ISOTROPIC TF, SYMMETRY (P1 OR P-1), OT OF (5041)
             OT   INDICES, WEIGHT, FO, FC. FOR SYMMETRY P1 ALSO  AC, BC, AC/FC, BC/FC
                  EXTRA CARD  SCL (CALCULATED BY LS OF WEIGHTED FO-FC)
  177 650         SF AT 3  BROWN,JENSEN,LINGFTR*,STEWART/400X,420X/                LWA
             4000,4010/PARAMETER LOADING AND GENERATION PROGRAMS
             4001,4002/AT 2 ONLY WITH PARTIAL CONTRIB, MAX 100 ATOMS, 8 TYPES
             4201/AT 3 AND DISPERSN CORRECTIONS, MAX 100 ATOMS, 8 TYPES, ANY ORDER
  176 650         SF AT 3  SHIONO            TRICL,MONOCL,ORTHO,8 KIND, 50 UNIQ    LWA
             ISOTROPIC OR ANISOTROPIC. BASIC 650 OR WITH IXA OR WITH IXA, IAS.
  235 650         STAT     ALDEN,STOUT*      1420/UNIT SF, WILSON + HPR STAT TESTS LPA
 5031 650         STAT     BARTL               VARIANCE'TEST                       LWA
             COMPUTES SPECIFIC VARIANCE FROM F SQUARED,PUT ON ABSOLUTE SCALE
             BY WILSON TEST.
 5030 650         STAT     BARTL               N(Z)'TEST                           LWA
             PREPARES DATA FOR N(Z)'PLOT.
  219 650         TF       STEWART/LINGFTR*/4500 TF CORR BY LST SQ                 LWA
             CALCULATES LST SQ ISOTROPIC TEMP FACTOR AND SCALE FACTOR CORRECT FROM
             OUTPUT OF STRUCT FACTOR PROGRAMS (IUCR 177) FOR FO RESCALING (IUCR 218)
  326 650         TF       SHIONO              ANISOTROPIC TF FROM PGM 174,176     LWA
             BASIC 650. USE OUTPUT OF PGM 174
```

328	650	TF UNIF	SHIONO	WILSON PLOT, UNITARY SF	LPA

650 WITH IXA,IAS, ON-LINE 407.

233 704   DIR   WOOLFSON   MULTIPLE SOLUTION DIRECT METHOD
DIRECT SIGN DETERMINATION USING METHOD OF STRUCTURE INVARIANTS AND Z-TEST
-- MULTIPLE SOLUTION RESULTS DISPLAYED AS FOURIER MAPS. WRITTEN IN
FORTRAN.

5025 704 8K   FR 2,3 HILDEB,HAHN*   GENER VERS BL GRID VARIABLE   LWA
CALCULATES SECTIONS PARALLEL A,B IN HEIGHTS Z. SIZE AND DENSITY OF
GRID IN A,B AND Z VALUES MAY BE SPECIFIED BY USER. C AND A BY SENSE
SWITCH. ANY SYM.IT HALF REC LATT, ONE CARD PER HKL COMBINATION, SORTED
ON L,K,H. IT AND OT VIA TAPE.

118 704 8K   FR 2 3 SLY,SHOEMAKER* MIFR1/GENL,FAST,BL,GRID N/120   LWA
SAP PROGRAM BEEVERS-LIPSON PRINCIPLE FAST TABLE LOOK-UP. APPLICABLE
TO ANY SPACE GROUP THRU ORTHORHOMBIC BY SELECTION OF SYMMETRY CARDS.
INPUT IS ONE CARD EACH UNIQUE REFLECTION. CARDS ARE SORTED ON INDICES
AND PUT ON TAPE. STRAIGHT OR DIFFERENCE FOURIER POSSIBLE. VARIABLE
FORMAT OUTPUT SHEETS PERMITS DIRECT CONTOURING. CALC UP TO 100,000
TERM POINTS PER SEC ON 32K 704. OPERATES ON 32K 709 WITH COMPATIBILITY.
HAS BEEN REWRITTEN FOR BIM 709/7090 (PGM 357).

301 704 4K   LC   MOZZI,NEWELL   RM1/THRU MONO, STD CALIBRATED. LS DETN LWA
DETERMINES LATTICE CONSTANTS FOR MONOCLINIC AND HIGHER SYMMETRY SYSTEMS
BY LEAST SQUARES METHOD UTILIZING AN ITERATIVE PROCEDURE9 DESIGNED FOR
DIFFRACTOMETER MEASUREMENTS CALIBRATED BY USING A STANDARD AND THEREFORE
NO EXTRAPOLATION FUNCTIONS ARE INCLUDED9 COMPUTES D VALUES FOR REQUESTED
(HKL) SETS.

4002 704FT28K 2DIRC WOOLFSON,SPARKS   LWM
INPUT DATA -- LC AND SF OR INTENSITIES. OUTPUT - FR FOR ACCEPTABLE SETS
OR SIGNS. 7 PACKS OF CARDS RUN CONSECUTIVELY. REQUIRES OPERATOR INTER-
VENTION AT SOME POUNTS. USES PRINCIPLES DESCRIBED IN ACTA CRYST 10,116/
11,277/11,393.

397 7040-44 FR3 HARRIS,D.R.   ERFR3/MAP   LPA
SUBSTANTIAL REVISION OF ERFR2 FOURIER OF SLY-SHOEMAKER-VAN DEN HENDE,
BASICALLY A FAP TO MAP CONVERSION. CHANGES INCLUDE SOME ADDITIONAL OP-
TIONS AND TAPE ASSIGNMENT CHANGES WITHOUT REASSEMBLY.

383 7070 AUC COORD SCHAPIRO   GENERATES ADDTL AT COORDS
381 7070 AUC FTN CLAY SHIONO   MCEWEN FT FOR LAYER STRUCTURE(CLAY) LWA
FOR CLAY MINERALS. REF. KOLLOID Z.,149,96(1956).

511 7070 AUC FR2 A CHU,MCMULLAN   MONOCLINIC A OR C PROJECTION LWA
USE OUTPUT OF PGM.330. 10K, 2 CHNL. ON-LINE 7500,7400 OR TAPE-1401

512 7070 AUC FR3 A CHU,MCMULLAN   MONOCLINIC CENT/NONCENT. LWA
USE OUTPUT OF PGM330. 10K,2 CHNL. ON-LINE 7500,7400 OR TAPE-1401 SYSTEM
NUMERIC OR ALPHABETIC OUTPUT FOR DIRECT PLOTTING

331 7070 ID CHU   DIST.ANGLE,S.D.OF ANGLE, ANY SYM. LWA
FORTRAN, 10K(5K), 7400, 7500

332 7070 ID CHU   INTER-,INTRA-MOLECULAR DIST.ANY SYM LWA
FORTRAN, 10K(5K), 7400, 7500

382 7070 FTN PLANE CHU   ANY SYM. 50 ATOMS MAX. LWA
330 7070 SF 3 SHIONO   TRICL,MONOCL,ORTHO LWS
10K(5K) CORE,7400,7500,2 CHNL 2 TAPE. INDIV ISO TF. 13 KIND,1500 ATOMS
AVAILABLE THROUGH GUIDE

510 7070 AUC SF3 AT SHIONO   SFIIA/ANY SYM. 460 ATOMS LWA
10K, 2 TAPES(2 CHNL.)CAN BE USED WITH ON-LINE 7400,7500 OR TAPE-1401
SYSTEM WITHOUT CHANGE.

338 7090 ABS BURNHAM   GNABS/CORR FOR XLS OF ARBITRARY SHAPE LWA
FORTRAN2, AND FAP (NO IT-OT), SUBROUTINES FOR EQUI-INCL WEIS GEOMETRY.
EASILY MODOFIED FOR GONIOSTAT GEOMETRY. 6010 + 13482 (COMMON) LOCATIONS.
CHOICE OF 64, 216, 512 INTEGRATION POINTS. LIMITS - 25 XTL FACES (NO
REENTRANT ANGLES), 25 RECIPROCAL LATTICE LEVELS.

5065 7090 FT2 ABS SCHULTZERHONHOF BN-ABS/ABS   LWA
MODIFIED VERSION OF BUSING OR-ABS (PGM 362). IT FROM BN-ST1, OT FOR
BN-ST2. RUNS INDEPENDENTLY OR WITH BN-X-64.

8018 7094 CONTOUR ANZENHOFER   PLOT OF FOURIER MAPS. LWA
PREPARES A MAGNETIC INPUT TAPE FOR AN AUTOMATIC PLOTTING DEVICE. PARTS OF
THE PROGRAM THAT DEPEND ON THE PLOTTER USED ARE INDICATED.

519 7090 45K CONTOUR CHERIN,MARTIN   PLOTS CONTOURS,LOCATES MAXIMA MPA
PROGRAM PLOTS THE RESULTS OF A FOURIER SYNTHESIS IN THE FORM OF CONTOURS.

```
       THE POSITIONS OF THE MAXIMA ARE DETERMINED    PLOTTED. PROGRAM REQUIRES
       40757 OCTAL STORAGE POSITIONS.WRITTEN IN FORTRAN
  368 7090      CONTOUR   VAN DEN HENDE   ERSCP/OT ERFR2 TO TAPE FOR SCOPE    NNX
       TAKES THE BINARY OT FROM ERFR2 AND PRODUCES A BINARY TAPE WHICH CAN BE
       USED TO PRODUCE CONTOUR MAPS ON AN OSCILLOSCOPE.
   49 709,7094 D H Q L*  WOLTEN         CALC ABOVE +THETA FROM LC,FT2,9PROGS  LWA
       2 OT LISTS, ORDERED ON H AND ON D
 5058 7090 FT2 D,P       KASPER,SCHMDMT* DWPULV/D,THETA FROM P                LWA
       DWPULV CALCULATES SIN,D,1/D**2,1/D AND CORRECTIONS BY K-ALPHA 1-2
       SPLITTING FOR POWDER FILMS ACCORDING TO THE ASYMMETRIC METHOD. RELIABLE
       REFLEXES ARE DISTINGUISHED AND USED TO CALCULATE THE LENGTH OF THE FILM.
  406 709,7094 D Q       WOLTEN         TABLES D,Q VS 2THETA INCR 0.01DEG     LWA
 8017 7094      DIR      ANZENHOFER     HARKER-KASPER INEQUALITIES            LPA
       INEQUALITIES OF THE TYPE 1-6 (SEE AC 3,436) CAN BE EVALUATED.THE PROGRAM
       IS USED AS PART OF A SYSTEM FOR APPLICATION OF DIRECT METHODS.
  475 7094 32K DP        CETLIN,B.B.    CALCS SF AND STD                      LPA
       FORTRAN PROGRAM TO COMPUTE AVERAGE VALUES FOR EACH STRUCTURE FACTOR
       WITH OBJECTIVE STANDARD DEVIATIONS.
  303 709      DP        PALENIK        14701/ PREP BIN DATA TAPE             LNA
       PREP BINARY DATA TAPE FOR INPUT INTO ALL PROGRAMS THREE POINT INTERPOLA-
       TION FOR SCATTERING FACTOR CAN ALSO BE USED TO CORRECT WEISSENBERG DATA
       ANY AXIS OR LEVEL
 5063 7090 FT2 DP        SCHULTZERHONHOF BN-X-64/SF,LP,ABS,SCL,FR,DF,LS FM,R,S LWA
       SYSTEM FOR RUNNING CRYSTALLOGRAPHIC PROGRAMS TOGETHER.(CHAIN) CONTAINS
       PGMS.BN-ST1,BN-ABS,BN-ST2,BN-LSQ,BN-FSY AND SOME SMALL PGMS. FOR DP.
       ALL IT-OT CARDS OR TAPE. EACH PROGRAM MAY ALSO BE RUN INDEPENDENTLY.
       COMPARE ALSO DESCRIPTIONS OF BN-ST1,BN-ABS,BN-ST2,BN-FSY + BN-LSQ.
 5066 7090 FT2 DP,SCL    SCHULTZERHONHOF BN-ST2/DP,SCL BY WILSON STAT OR SPEC. LWA
       IT FROM BN-ST1 AND BN-ABS.CORRECTS SF WITH ABS FROM BN-ABS(MAY BE OMITTED
       BY SWITCH),CALCULATES SCL BY SUBROUTINES(WILSON-STAT OR SYMMETRY). OT FOR
       BN-FSY FOR PATTERSON AND FOR BN-LSQ. MAY ALSO BE RUN WITH BN-X-64.
 8019 7094      EL       DALLINGA       LEAST SQUARES GASDIFFR. INTENSITIES   LWA
       REFINEMENT OF MOLECULAR COORDINATES AND TEMPERATURE FACTORS.TWO TYPES OF
       CONSTRAINTS CAN BE USED A) LINEAR EXPRESSIONS INVOLVING 1,2 OR 3
       INTERATOMIC DISTANCES, B) SCALE FACTOR AND TEMP.FACTORS MAY BE KEPT
       CONSTANT (SEE REC.TRAV.CHIM.83,789(1964)
 7027 7094 II FM E S     TOURNARIE      310D/OPTIMAL LINEAR AJUSTMENTS        LWA
  322 7090/709 FPK       IBERS          27 PT. LS 10 PARAMETER GAUSSIAN FIT   LNA
       PROGRAM TAKES 27 POINT INPUT FROM MAP AND FITS THESE DATA WITH A 10 PARA/
       METER GAUSSIAN FUNCTION. AXES NEED NOT BE ORTHOGONAL NOR INTERVALS EQUAL.
       OUTPUT INCLUDES PEAK CENTER, HEIGHT, GAUSSIAN CONSTANTS. 7090 TIME IS 3
       SECONDS/PEAK.  WRITE UP AVAILABLE IN FORM OF FORTRAN LISTING WITH COMMENT
       CARDS.  COMPILATION IS LEFT TO USER.
  491 7090      FPK      SANBORN        LINE PROF ANAL                        LPA
       LINE PROFILE FOURIER ANALYSIS. PROGRAMS IN FORTRAN BY KELLER AND SEGMUL-
       LER (REV SCI INST 34,684-8(1963)) FOR 1620 MODIFIED AND ADAPTED TO 7090.
       INPUT IS STEP SCAN FIXED COUNT DATA. CORRECTIONS APPLIED FOR BACKGROUND,
       ATOMIC SCAT FACTOR, LP FACT, ABSORPTION. OUTPUTS ARE CENTROID, VARIANCE,
       FOURIER COEFFICIENTS.
  128 7094      FR 2 3   BRYDEN         2 AND 3 DIM FOURIER SUM.              LPA
       COMPUTES 2 AND 3 DIM. FOURIER SUMMATIONS BY A TABLE LOOK-UP PROCESS.
       THREE TAPE UNITS ARE USED TO SORT AND STORE DATA.
  391 7090      FR3 CNTR LEVY,ELLISON*  XFOUR/FOURIER SYNTH, CONTOURS PK POS  LPA
       MINOR MODIFICATION OF A. ZALKIN PROG. FORDAP. OPTIONS FOR FOURIER,
       PATTERSON, DIFFERENCE AND PARTIAL DIFFERENCE MAPS. SEARCH FOR POSITIVE
       AND NEGATIVE PEAK POSITIONS. CONTOUR MAPS ON CATHODE RAY PLOTTER.
       MOSTLY FORTRAN 2, PARTLY FAP.
  357 7090 32K FR 3 2    SLY,SHO,HENDE* ERFR2/GENL,FAST,BL,120 GRID (MIFR1)   LWA
       MODIFICATION OF MIFR1 (PGM 118) FOR IBM 709/7090, 32K REQUIRED. GENERALLY
       SIMILAR, BUT WITH SOME ADDED CONVENIENCE FEATURES. FAST TABLE LOOKUP.
       ANY SPACE GROUP TRICLINIC MONOCLINIC ORTHORHOMBIC BY SELECTION OF
       SYMMETRY CARDS.  INPUT IS ONE CARD EACH UNIQUE REFLECTION.  CARDS ARE
       SORTED ON INDICES AND PUT ON TAPE.  STRAIGHT OR DIFFERENCE FOURIER
       POSSIBLE.  VARIABLE FORMAT OUTPUT SHEETS PERMIT DIRECT CONTOURING.
       CALCULATES UP TO 300,000 TERM POINTS PER SECOND.  CAN BE RUN ON
       IB FORTRAN MONITOR SYSTEM.
```

THE SUPPLY OF ERFR2 MANUALS IS EXHAUSTED, BUT A COPY OF THE MANUAL
HAS BEEN DEPOSITED AS ADI DOCUMENT NO 8933 WITH THE AMERICAN DOCUMENTA-
TION INSTITUTE, U. S. LIBRARY OF CONGRESS, WASHINGTON 25, D. C.
A COPY MAY BE OBTAINED BY CITING THE DOCUMENT NUMBER AND REMITTING
$17.50 FOR PHOTOPRINTS OR $5.50 FOR MICROFILM. ADVANCE PAYMENT IS
REQUIRED. MAKE CHECKS OR MONEY ORDERS PAYABLE TO **CHIEF, PHOTO-
DUPLICATION SERVICE, LIBRARY OF CONGRESS**.
NOTE -- AN UPDATED VERSION OF THE PROGRAM FOR THE IBM 7094 IS
MIFR2A (Q. V.).

534 7094   FR 3 2   SHOEMAKER,KATZ   MIFR2A/ GENERAL FOURIER   LPA
   MODIFIED VERSION OF SLY, SHOEMAKER, VAN DEN HENDE ERFR2 7090 CRYSTAL-
   LOGRAPHIC FOURIER PROGRAM (Q. V.). CONTAINS ADDITIONALLY A RESCALING
   FEATURE TO AUTOMATICALLY CORRECT TOO HIGHLY SCALED INPUT DATA.
   PROGRAM IS IN FAP LANGUAGE, SOURCE PROGRAM AVAILABLE BY SENDING
   MAGNETIC TAPE TO D. P. SHOEMAKER, M. I. T.

394 7090 FT2 FR3 SPEC BROWN G M   COMFO/3D COMPOSITE, SQ. MESH   LPA
   COMPUTES PARALLEL SECTIONS THRU CENTERS OF ATOMS ON SQUARE MESH. SEC-
   TIONS MAY HAVE ANY ORIENTATION. INPUT TAPE FROM XFLS (AC 389) OR BMFLS
   (AC 390). FOR SPECIAL PURPOSES, NOT GENERAL USE (SLOW).

408 7094   FR SPEC DONNAY,TAKEDA   JH-ORTFR/DIRECT PRINT UNDEFORMED MAP LPA
   CALCULATES FR SERIES AT THOSE POINTS WHERE COMPUTER PRINTS THEM,
   WE CAN DRAW CONTOURS ON DIRECT PRINT-OUT. WEITTEN IN FORTRAN FOR
   MONOCL. C, HOL PROJECTION AND SECTION, BUT BE EASILY GENERALIZED
   DESCRIPTION TO APPEAR IN *SCIENCE*.

453 709 FTN FT   OCKEN*WAGNER   RADIAL DISTRIBUTION FUNCTION   LWA
   COMPUTES RDF FROM MEASURED X-RAY SCATTERING OF LIQUIDS

503 709 FT2 FT   OCKEN/YCC   RDF OF LIQUIDS   LWA
   X-RAY DATA CORRECTED FOR POLARIZATION. RESULTS NORMALIZED TO ABSOLUTE
   UNITS. SI(S) FUNCTION GENERATED. RDF DETERMINED BY CALCULATION OF THE
   FOURIER SINE COEFFICIENTS OF SI(S) USING FILON'S METHOD. STORAGE, 9K.

5067 7090 FT2 FS,DF   SCHULTZERHONHOF BN-FSY/FS,DF,PATTERSON. BL.   LWA
   SYSTEM OF FS PGMS.ALL SPACEGROUPS.RUNS INDEPENDENTLY OR WITH BN-X-64.
   IT POSSIBLE FROM BN-ST2 FOR PATTERSON, FROM BN-LSQ FOR FS OR DF.
   REFLECTIONS NEED NOT TO BE SORTED.BLK.

412 IBM7094 GENERAL STEWART*HIGH   X-RAY-63/SYSTEM FOR IBM 709-7094-(DC) LWA
   SEE ALSO UNDER X-RAY-63 ENTRY IN LIST FOR MORE DETAIL

363 7090 FT2 ID E   BUSING*LEVY   ORFFE/FORTRAN FUNCTION AND ERROR   LWA
   USED INDEPENDENTLY OR WITH ORFLS. COMPUTES THE FOLLOWING FUNCTIONS OF
   CRYSTAL PARAMETERS WITH THEIR STANDARD ERRORS- DISTANCES, ANGLES,
   DIHEDRAL ANGLES, RMS PRINCIPAL THERMAL DISPLACEMENTS, ANGLE BETWEEN
   PRINCIPAL AXIS AND VECTOR DEFINED BY USER, RMS RADIAL THERMAL DISPLACE-
   MENT, DISTANCE CORRECTED FOR THERMAL MOTION. WRITTEN IN FORTRAN 2.

473 7094 32K IT DP   ABRAHAMS,CETLIN PEXRAD INPUT/AUTO DIFFR CONTROL TAPE LPA
   FORTRAN PROGRAM TO ARRANGE VALUES OF H,K,AND L,BY LTH LAYERS,
   IN OPTIMUM ORDER, TO CONTROL CRYSTAL AND COUNTER SGTTINGS ON BELL
   TEL. LABS. PROGRAMMED ELECTRONIC X-RAY AUTOMATIC DIFFRACTOMETER
   (PEXRAD). TO BE USED IN CONJUNCTION WITH PEXRAD OUTPUT AND F**2 PROG.

515 7090   IT/OT/DP ABRAHAMS/CETLIN PEXRAD AUTOMATIC DIFFRACTOMETER PRGM LWA
   FORTRAN PROGRAM TO CONTROL OPERATION OF AUTOMATIC DIFFRACTOMETER + CHECK
   ITS PERFORMANCE. STATISTICAL ANALYSIS OF EQUIVALENT STRUCTURE FACTORS
   GIVES MEAN STRUCTURE FACTOR AND STANDARD DEVIATION .

451 7090 LC   BURNHAM   LCLSQ/LS REFINEMENT OF LC   LWA
   FORTRAN2, + FAP (NO IT-OT) SUBROUTINES, 5173 + 1900 (COMMON)
   LOCATIONS (+ USER SUBROUTINES). LS REFINEMENT OF LC FOR ANY XTL SYSTEM.
   USES INDEXED SINGLE-XTL OR POWDER DATA. ALLOWS REFINEMENT OF 9 OR LESS
   SYSTEMATIC CORRECTION TERMS, E.G. ABSORPTION, SHRINKAGE, ETC. USER
   PROGRAMS TRIGONOMETRIC FORM OF CORRECTION TERMS WITH POSSIBILITY OF 5
   DIFFERENT FORMS. PROVISIONS FOR WEIGHTING.

518 7094   LC BY LS2 OTTE,ESQUIVEL HEXAG LC BY NEW LS-2 FROM D-S PATTERN LWA
   A NEW LS METHOD FOR PRECISE DETERMINATION OF HEXAGONAL LC FROM
   D-S(DEBYE-SCHERRER)PATTERNS. USES REITERATIVE PROCESS AND CORRECTION
   FACTOR WHICH,UNLIKE COHEN S (REV SCI INSTR,1935)VARIES LINEARLY WITH
   EXTRAPOLATION FN. PROVISION MADE FOR VARYING XRAY WAVE LENGTH
   AND WEIGHTING FACTOR.

47 7094 FT2 LC AND E VOGEL,KEMPTER* HERTA-2/PREC REFINMT OF LATTICE PARAM LWA
   A CONVERGENT ITERATIVE TECHNIQUE (R.E.VOGEL AND C.P.KEMPTER,ACTA CRYST.

14,1130(1961)),WITH EXACT STATISTICAL WEIGHTS ASSIGNED TO THE OBSERVED
POINTS IS USED TO DETERMINE LATTICE PARAMETERS AND THEIR STANDARD DEVI-
ATIONS FOR THE CUBIC,HEXAGONAL,RHOMBOHEDRAL(ON HEXAGONAL
AXES),TETRAGONAL,ORTHORHOMBIC, AND MONOCLINIC SYSTEMS FROM DEBYE-SCHERRER,SYMMETRICAL
BA3K REFLECTION,OR DIFFRACTOMETER DATA (FOUR EXTRAPOLATION FUNCTIONS).
130 7094   LP   BRYDEN   CORR. EQUIINCL. WEIS. INT.   LPA
CORRECTS EQUIINCLINATION WEISSENBERG INTENSITIES FOR LP FACTORS, AND FOR
UPPER LAYER LINE SPOT EXTENSION BY THE METHOD OF PHILLIPS.
310 709   LP   PALENIK   14708/ANY SPACE GROUP   LNA
320 7090 32K LP ABS H VAN DEN HENDE   ERBR2/LP AND ABSN CORR,SHARPENING   LPA
ERBR2 CORRECTS FOR LP AND ABSORPTION. NORMAL-BEAM AND EQUI-INCLINATION.
CYLINDRICAL OR SPHERICAL SPECIMENS. ALSO SHARPENING FOR B POSSIBLE.
GENERATION OF INDICES OPTIONAL, ALSO ADDITIONAL OUTPUT IN FORMAT OF
SLY-SHOEMAKER FOURIER PROGRAM. MAY BE USED WITH 704 ALSO WITH MINOR
CHANGES.
513 7090FT2   LP   SMITH D K   WEIS OR PREC LP CORR   LPA
ALSO SCALES INTERSECTING SECTIONS BY LS FIT AND GIVES OT FOR SF CALC.
8527 7090   LS   ASHIDA   LSQALLTB/BLOCK-DM FORTRAN 2 AND 4   LWA
BLOCK-DM LS. AT,E,DISPERSION. 4*4 OR 9*9 FOR EACH ATOM. 2*2 FOR SCL AND
OVERALL TF. ANY SPACE GROUP. 60 ATOMS. 5000 REFLECTIONS.
8530 7090   LS   ASHIDA   LSQAL/BLOCK-DM. FORTRAN 2 AND 4   LWA
BLOCK-DM LS. AT,E,DISPERSION. 4*4 OR 9*9 FOR EACH ATOM, 2*2 FOR SCL
AND OVERALL TF. ANY SPACE GROUP. 60 ATOMS. 5000 REFLECTIONS.
361 7090 FT2 LS   BUSING*LEVY   ORGLS/GENL LS TO FIT ARBITRARY FUNCT. LWA
PERFORMS LEAST SQUARES FIT OF ANY FUNCTION DEFINED BY USERS SUBROUTINE.
DERIVATIVE CALCULATION PROGRAMMED OR NUMERICAL. OBSERVATIONS WEIGHTED
UNITY OR INDIVIDUALLY. TRIAL PARAMETERS TO BE VARIED ARE SPECIFIED FOR
EACH JOB. COMPARISON OF CALCULATED AND OBSERVED FUNCTION PUT OUT BEFORE
EACH CYCLE AND AFTER LAST. ADJUSTED PARAMETERS AND STANDARD ERRORS PUT
OUT AFTER EACH CYCLE. CORRELATION MATRIX PUT OUT. MATRIX INVERTER IS FAP
FOR 704, 709, OR 7090. VERSION ENTIRELY IN FORTRAN IS ALSO AVAILABLE.
5068 7090 FT2 LS   SCHULTZERHONHOF   BN-LSQ/LS,FM,R,TF,SIGN   LWA
MODIFIED VERSION OF SCHERINGER LS. IT FROM BN-ST2. OT FOR BN-FSY(FS,DF).
RUNS INDEPENDENTLY OR WITH BN-X-64.
360 7090 FT2 LS AT FM BUSING*LEVY   ORFLS/FORTRAN LEAST SQUARES, SF,FM,AT LWA
STRUCTURE FACTOR LEAST SQUARES. ANY SPACE GROUP. VARIABLES ARE SCALE
FACTORS, NEUTRON SCATTERING FACTORS, ATOM MULTIPLIERS, ISOTROPIC OR
ANISOTROPIC TEMPERATURE FACTOR COEFFICIENTS. FULL MATRIX LEAST SQUARES.
GENERAL OR UNIT WEIGHTS. REFINEMENT BASED ON F OR F SQUARED. ARBITRARY
SELECTION OF PARAMETERS VARIED. 32K MEMORY USUALLY REQUIRED. USES NO
TAPES FOR STORAGE. MATRIX INVERTER IS FAP FOR 704, 709, OR 7090.
VERSION ENTIRELY IN FORTRAN IS ALSO AVAILABLE.
399 7090/94 LS BM AT VAN DEN HENDE   LLXR6/BLOCK DIAGONAL LEAST-SQUARES   LWA
LLXR6 ALLOWS ANISO AND ISO TEMP FACTORS, GENERAL WEIGHTING AS WELL
AS DIFFERENTIAL SYNTHESIS WEIGHTING. IT WILL VARY 10 PARAMETERS FOR 176
ATOMS SIMULTANEOUSLY FOR ALL 230 SPACE GROUPS. CAN BE USED FOR STRUCTURE
FACTOR CALC ALSO. SIMPLE CONTROLS ALLOW MORE THAN 20 OPTIONS. NEUTRON
DATA ARE ALLOWED, ANOMALOUS SCATTERING IS TAKEN CARE OF. OUTPUT FOR
ERFR2 FOURIER PROGRAM MAY BE WRITTEN. SUPERSEDES ERBR1. VERY FAST.
384 7090   LS FM   GANTZEL,SPARKS,TRUEBLOOD/UCLALS1/REVN OF PGM 228
FULL-MATRIX LEAST SQUARES REFINEMENT OF SUITABLE TRIAL CRYSTAL
STRUCTURES. REFINES UP TO 137 UNIQUE PARAMETERS, WITH UP TO 2100 UNIQUE
F(HKL). EVERYTHING IS STORED IN HIGH-SPEED MEMORY. A CONSIDERABLE
REVISION OF PROGRAM 228.
308 709   LS FM   PALENIK   14706/ ANY ACENTRIC SPACE GROUP   LNA
FULL MATRIX LEAST SQUARES ANY ACENTRIC SPACE GROUP USES UNIQUE DATA ONLY
LIMIT OF 30 ATOMS FOR SF PART AND 18 ATOMS WITH ANISOTROPIC TEMPERATURE
PARAMETERS FOR LEAST SQUARE PART RECYCLE CHANGING ONLY VECTOR OR
VECTOR AND MATRIX
309 709   LS FM   PALENIK   14707/ ANY CENTRIC SPACE GROUP   LNA
SAME AS 14706 EXCEPT FOR CENTRIC SPACE GROUPS
317 709   LS H FM GTZL,SPKS/TRBD*   UCLALS1/137 PAR,ISO OR AT, 2100 REFL LPA
UCLALS1/ ALL 2100 REFL DATA STORED IN MEMORY. NON-LIMITED ARE NO. OF
CYCLES, S. G., SPEC POSNS. ROOM TO ADD VARIOUS SUBPROGRAMS SUCH
AS DISTANCE-ANGLE, R BY CLASSES, DELETION, E.S.D., ETC. MODIFICN TO READ
IN AND PROCESS ONE REFL AT A TIME EASILY PERMITS INF DATA AND 152 PARAMS

```
4001  7090FT2   LS2 FMAT   CURTIS/GILMARTIN/LS2D/GENL 2D LS,FM,AT,N OR XRAY          LWA
        GENERAL 2-DIM SF + LS PGM, FULL MATRIX, ANISO. NEUTRON OR XRAY. INPUT
        PLANE GP NO, LC, LF, LIST OF HKF, PARAS, CONTROL CODE FOR EACH CYCLE.
        WEIGHT IS A FUNC OF F. OUTPUT PARAS, CHANGES, ERRORS, SF, AS REQD.
        WRITE-UP AERE-R3134 AMENDED BY CPN 52. SAME AS MERCURY PGM. NO. 29F
        (ENTRY NO. 4007).  UP TO 50 DISTINCT ATOMS, 100 PARAS REFINED, 1000
        PLANES, ON 32K 7090.
 311   709        M SQ DIS   PALENIK            14709/MEAN SQ. DISPLACEMENTS          LNA
 516   7090       OT/DP     BERNSTEIN,J.L.    ANOMALOUS DISPERSION PROGRAM            LWA
        FORTRAN PROGRAM CORRECTS OBSERVED STRUCTURE FACTORS OF A CENTRO-SYMMETRIC
        CRYSTAL FOR ANOMALOUS DISPERSION. PROGRAM USED WITH ONE OR TWO ATOMS
        DISPERSIVE. ONE CARD OBTAINED FOR EACH REFLECTION. CARDS PUNCHED IN
        FORMAT OF INPUT CARDS OF ORFLS
 474   7094 32K OT DP   ABRAHAMS,CETLIN PEXRAD OUTPUT/CALCS SF AND ERRORS            LPA
        FORTRAN PROGRAM TO INTEGRATE INTENSITY FROM BETA-ALPHA PROFILE,MAKE
        ALL CORRECTIONS AND COMPUTE STRUCTURE FACTORS,INDICATING ALL ERRORS.
 514   7090FT2  P       SMITH D K      POWD/ COMPUTES POWDER INTENSITIES             LWA
        INCLUDING LP, MULTIPLICITY, AND ABSORPTION FOR ANY SYMMETRY
 304   709       PATT SUM   PALENIK         14702/ 3D PATTERSON MONOCLINIC           LNA
        WILL PRODUCE A SHARPENED OR UNSHARPENED PATTERSON  WITH OR WITHOUT THE
        ORIGIN PEAK  3D LIMIT OF 2000 REFLECTIONS
 305   709       PATT SUM   PALENIK         14703/ 3D PATTERSON ORTHORHOMBIC         LNA
        SAME AS 14702 EXCEPT FOR ORTHORHOMBIC SPACE GROUPS
3041   7090 FT2 ROTF     ROSSMANN       GRF2/ROTATION FUNCTION*ALL SP.GRPS           LWA
        USES INTENSITIES TO FIND ANGULAR RELATIONSHIPS BETWEEN MOLECULES OR
        SUB UNITS WITHIN THE SAME ASYMETRIC UNIT OR IN DIFFERENT CRYSTAL
        MODIFICATIONS(ROSSMANN,BLOW ACTA 1962,VOL15,P.24)
7026   7094 II  S DP    TOURNARIE         308B/OPTIMAL SMOOTHING                     MWA
 404   709,7094 SF,ICA    WOLTEN         CALC FROM LC AND AT PAR FT2                 LWA
        OT SAME AS PROG 49 PLUS SF,ICA
 306   709       SF FR     PALENIK         14704/ACENTRIC THRU ORTHORHOMBIC          LNA
        CALC SF AND USE SIGNS FOR FOURIER CALC BASED ON SIMPLFIED FORMS IN
        INTL.TABLE USES ONLY THE UNIQUE INPUT ANY INTEGER VALUE OF 120 CAN BE
        USEDFOR FOURIER INTERVAL
 307   709       SF FR     PALENIK         14705/ CENTRIC THRU ORTHORHOMBIC          LNA
        SAME AS 14704 EXCEPT FOR CENTRIC SPACE GROUPS
8526   7090      SF FR     SASADA,ASHIDA    SFFR/SF,FR IN SUCCESSION FORTRAN2        LWA
        CALCULATES SF AND FR. APPLICABLE TO TRI,MONO,ORTHO.
        100 ATOMS, 3000 REFLECTIONS
 316   709       SF H FR   GANTZEL/TRBD*    UCLANC1/ SEE ABSTR FOR LIMITATIONS       LPA
        UCLANC1/ ANY SG, 100 ATOMS, ANISO OR ISO B. FOR FOURIER, EQUIV DATA GEN
        FOLLOWS SF (WILL NEED PATCHES FOR CERTAIN SG). MAX INDEX PROD LIMITED TO
        ABOUT 2700 (SOON 5500) BECAUSE INTERNAL SORTING. NO. OF 4-TERM SUMS (HKL
        WITH ALL SIGNS FOR KL) LIMITED TO 1500. WRITTEN FOR NON-CENTROSYM( EASY
        MODIF FOR 2400 CENTROSYM 4-TERM SUMS).
5064   7090 FT2 SF,LP    SCHULTZERHONHOF  BN-ST1/SF,LP,THETA                         LWA
        IT FROM WEIS OR COUNTER.CALCULATES SF,LP,FILM CORRECTIONS,THETA, OT FOR
        BN-ABS AND BN-ST2. MAY ALSO BE RUN WITH BN-X-64.
 291   709 FT2  SF 2     NORTH           SF2/2/FAST 2-DIMENSIONAL                    LWA
        2-DIM. STRUCTURE FACTORS, PLANE GROUPS 1 - 9, SYMMETRY SPECIFIED BY
        PLANE GROUP NO. + 2 SELECTOR SYMBOLS, UNLIMITED NO. OF ATOMS,
        ISOTROPIC TEMP. FACTORS
 315   709       SPC N    SHOEMAKER,C+D    PRPLOT/SMOOTH AND PLOT SPECROM DATA       LPA
        709 VERSION OF PGM 122
 393   7090 FT2 SPEC     BROWN G M        EDIT/F SQUARE TABLE FOR PUBLIC.            LPA
        MINIMIZES CUTTING AND PASTING.  SETS UP TABLE IN MULTICOLUMN FORMAT WITH
        NUMBERS OF LINES AND COLUMNS VARIABLE.  INPUT TAPE FROM XFLS (AC 389) OR
        BMFLS (AC 390).  FOR EXAMPLE, SEE R. D. ELLISON AND H. A. LEVY, ACTA
        CRYST. IN PRESS, 1965.
 480   7094      SPEC     DUCHAMP,MARSH*,ETAL/CRYRM/COMPLETE CRYSTALLO. SYSTEM       LWM
        CONTROL SYSTEM AND SUBPROGRAMS FOR INITIAL DATA REDUCTION(LP,SCALING,ETC)
        STRUCTURE-FACTOR LEAST SQUARES(SPACE GROUP SPECIFIC,ALL SPACE GROUPS,FM,
        BLOCK MATRIX,AT,ETC.),FOURIER(DIFFER,GEN.PLANE,2 OR 3,TO-SCALE MAPS),
        DISTANCES,ANGLES,L.S.PLANES,AUTOMATIC DIFFRACTOMETER CONTROL,LATTICE
        CONSTANT DETERMINATION,PUBLICATION LIST,TEMP. ELLIPSOIDS,STATISTICAL
        PHASING.  PROVIDES FOR VERY LARGE PROBLEMS.
```

```
230 709      SPEC     SPKS,BURKE/TRBD UCLAMO1/MOLEC ORIENTN,PGG,PGM,PG    LPA
    UCLAMO1/  EVALUATES A SORT OF R FOR A SPECIFIC ORIENTATION OF A MOLECULE
    OF KNOWN GEOMETRY TRANSLATED OVER A GIVEN PROJECTION. THREE ANGULAR PARAM
    TO BE SPECIFIED (FROM OTHER CONSIDERATIONS). MINIMA ARE PLAUSIBLE POSNS
231 709      SPEC     SPARKS/TRBD*    UCLAMOLS1/5 PAR LS OF 230 FOR PGG   LPA
    UCLAMOLS1/ REFINES THE FIVE APPROX PARAMS FROM 230 ONLY FOR PLANE GROUP
    PGG BY LS (THREE ANGLES, TWO POSITIONS)
405 709,7094 SPEC     WOLTEN          CALC INTERPLANAR ANGLES FROM LC,FT2 LWA
448 IBM7094  SPEC,DP  LEDLEY,DAYH,JSTEW/AUTOMATIC SCAN DIFF FILM TO CORE MEMNPA
    DIFFRACTION FILM OPTICAL DENSITIES ARE SCANNED AUTOMATICALLY INTO THE
    CORE MEMORY OF AN IBM 7094 USING THE =FIDAC= FLYING SPOT SCANNER. EACH
    DIFFRACTION SPOT IS COVERED BY SEVERAL HUNDRED SCANNER POINTS.
    INTEGRATED INTENSITY OF SPOTS IS OBTAINED BY FORTRAN AND FAP OPERA-
    TIONS ON THE DENSITIES OF SPOT AND BACKGROUND POINTS.
407 7094     SPEC LC  TAKEDA,DONNAY   JH-TRXL63/TRANSF. XL. SETTING       LPA
    TRANSFORMS CELL DIMENSION, SPACE-GROUP SYMBOL TO A CONVENTIONAL SETTING
    OF *CRYSTAL DATA*. RHOMBOH.-HEXAG. TRANSF., REORIENTATION OF ORTHOR.,
    CELL REDUCTION OF MONOCLI. AND TRICLI.. USES SAME INPUT FORMAT.
    WRITTEN IN FORTRAN AND FAP(SOME SUBROUTINE).
388 7090     STEREO   JOHNSON C K     ORTEP/THERMAL ELLIPSOID PLOT PROGRAM LWA
    UTILIZES PLOTTER SUCH AS CALCOMP TO PLOT BALL-AND-STICK CRYSTAL STRUCTURE
    FIGURES IN STEREO WITH THERMAL ELLIPSOIDS OR CIRCLES ON THE ATOMIC SITES.
    MAIN PROGRAM IS IN FORTRAN II. IBM 7090 OR CDC 1604 MACHINE ORIENTED PLOT
    ROUTINES ARE LIBRARY TYPE WITH MINOR MODIFICATIONS. 32K MACHINE REQUIRED.
232 709      TF AT    CLTR,GTZL/TRBD* UCLATO1/RIGID BODY, TRANS AND LIBR  LPA
    UCLATO1/ OK FOR ANY AXIAL SYSTEM.  USES CRUICKSHANK RIGID BODY APPROX
    INPUT ANISOT B'S AND ATOMIC POSNS. OUTPUT PARAMS INDIVID ATOMIC VIBR
    ELLIPSOIDS, T, OMEGA, AND ESD'S, AND DELTA U(IJ)
```

ACCESSION NUMBER INDEX

FOLLOWING EACH ACCESSION NUMBER THE FIRST FIVE CHARACTERS OF THE MACHINE
ABBREVIATION AND THE FIRST TWO CHARACTERS OF THE FUNCTION ABBREVIATION ARE
GIVEN, TO FACILITATE FINDING THE PROGRAM IN THE MAIN LIST.

FROM 1ST ED.	260 220 DF	363 7090 ID	441 X-RAY DP
1 LGP30 DP	261 220 ID	368 7090 CO	442 X-RAY DP
2 LGP30 FR	262 220 LF	369 1620 SF	443 X-RAY DP
3 LGP30 FR	263 220 LP	370 1620 FR	444 X-RAY SP
34 650 AB	264 220 H	371 1620T FR	445 X-RAY SP
47 7094 LC	265 220 FR	381 7070 FT	446 1604 FR
49 709,7 D	266 205 FR	382 7070 PL	447 1604 FR
111 205 H	267 205 EL	384 7090 LS	448 IBM70 SP
112 205 ID	268 220 LS		449 FT2 CO
113 205 LP	269 205 EL	NEW, 2ND ED.	450 FORTR AB
114 205 FR	277 1620 D	385 FORTR N	451 7090 LC
115 205 LP	278 1620 LC	386 FORTR LC	452 FORTR PA
116 220 LF	279 1620 D	387 1604 ST	453 709 F FT
118 704 8 FR	280 1620 H	388 7090 ST	454 1620 DP
120 FORTR DP	281 1620 H	389 FORTR LS	455 1620 LF
128 709 FR	283 1620 D	390 FORTR LS	456 1620 DF
130 709 LP	284 1620 D	391 7090 FR	457 1620 SF
133 650 PA	285 1620 D	392 1604A SP	458 1620 E
170 650 FR	287 1620 D	393 7090 SP	459 1620 ID
171 650 FR	288 1620 LC	394 7090 FR	460 1620 ID
172 650 FR	289 1620 LC	395 FORTR FT	461 1620 D
174 650 FR	290 650 H	396 FORTR ID	462 1620 SP
176 650 SF	291 709 F SF	397 7040- FR	463 FTN 1 LC
177 650 SF	292 1620 AB	398 FORTR SF	464 FTN 1 SP
180 650 DP	294 1620 AT	399 7090/ LS	465 FTN 1 DP
181 650 DP	297 220 SP	400 1620 FR	466 FTN 1 AB
182 650 PL	298 220 DP	401 1620 FR	467 FTN 1 LP
183 650 HY	300 220 FR	402 1620 SF	468 FTN 1 ST
184 650 ID	301 704 4 LC	403 1620 SF	469 FTN 1 LS
185 650 DP	302 650 FP	404 709,7 SF	470 FTN 1 ID
198 650 FR	303 709 DP	405 709,7 SP	471 FTN 1 ID
218 650 DP	304 709 PA	406 709,7 D	472 FTN 1 DE
219 650 TF	305 709 PA	407 7094 SP	473 7094 IT
220 650 ID	306 709 SF	408 7094 FR	474 7094 OT
221 650 ID	307 709 SF	409 ALGOL ID	475 7094 DP
225 FORTR AB	308 709 LS	410 ALGOL LS	476 FORTR DP
226 FORTR LC	309 709 LS	411 FORT2 GE	477 FORTR LS
230 709 SP	310 709 LP	412 IBM70 GE	478 FORTR FR
231 709 SP	311 709 M	413 X-RAY GE	479 FORTR ID
232 709 TF	312 1620, D	414 X-RAY SF	480 7094 SP
233 704 DI	313 FORTR LS	415 X-RAY LS	481 FORTR PI
235 650 ST	314 FORTR SP	416 X-RAY LS	482 FORTR PI
236 ALGOL D	315 709 SP	417 X-RAY LS	483 FORTR PI
237 ALGOL SF	316 709 SF	418 X-RAY FR	484 FORTR PI
238 ALGOL LP	317 709 LS	419 X-RAY CO	485 1620 FR
239 ALGOL S	320 7090 LP	420 X-RAY PK	486 1620 FR
240 ALGOL H	322 7090/ FP	421 X-RAY DP	487 1620 SF
241 ALGOL LC	324 650 DP	422 X-RAY SP	488 1620 LF
242 ALGOL TF	325 650 E	423 X-RAY DI	489 1620 LP
243 ALGOL LF	326 650 TF	424 X-RAY ST	490 1620 ID
244 ALGOL LP	327 650 LF	425 X-RAY SP	491 7090 FP
245 ALGOL S	328 650 TF	426 X-RAY ST	492 1105 SF
246 ALGOL PR	329 650 DP	427 X-RAY LC	493 1105 FR
247 ALGOL E	330 7070 SF	428 X-RAY TH	494 1105 FR
248 ALGOL DF	331 7070 ID	429 X-RAY WE	495 1105 LP
249 ALGOL DF	332 7070 ID	430 X-RAY SP	496 FORTR ST
250 ALGOL DF	338 709 3 AB	431 X-RAY FR	497 1620- FR
251 ALGOL DF	344 FORTR FR	432 X-RAY PA	498 1620- LP
252 ALGOL LS	345 FORTR N,	433 X-RAY ID	499 1620- ID
253 ALGOL LS	354 1620T FR	434 X-RAY ID	500 FORTR SP
254 ALGOL SF	355 1620T SP	435 X-RAY ST	501 FORTR LC
255 ALGOL SF	356 650 FR	436 X-RAY LS	502 BALGO TH
256 ALGOL FR	357 7090 FR	437 X-RAY SP	503 709 F FT
257 ALGOL SC	360 7090 LS	438 X-RAY DP	504 FORTR FR
258 ALGOL SC	361 7090 LS	439 X-RAY DP	505 FORTR SP
259 ALGOL SC	362 FORTR AB	440 X-RAY DP	506 FORTR ID

507 FORTR LP	3050 ATLAS SF	4042 MERCU TF	5066 7090 DP
508 650 FR	3051 ATLAS ID	4043 MERCU GE	5067 7090 FS
509 650 LP	3052 ATLAS CO	4044 MERCU TF	5068 7090 LS
510 7070 SF	3053 ATLAS 2C	4045 MERCU ID	5069 Z23 ST
511 7070 FR	3054 ATLAS SF	4046 MERCU GE	5070 Z23 FT
512 7070 FR	3055 ATLAS RO	4047 MERCU GE	5071 Z23 FO
513 7090F LP	3056 FORTR LS	4048 MERCU PK	5072 ALGOL P
514 7090F P	3057 ATL F LS	4049 DEUCE FR	5073 ER 56 AB
515 7090 IT	3058 ATL F FR		5074 FORTR PR
516 7090 OT	3059 ATL F DP	5001 ER 56 TH	
517 B 500 LS	3060 ATL F SF	5002 ER 56 TH	FROM 1ST ED.
518 7094 C	3061 ATL F PA	5003 ER 56 LC	6001 WGMAT LP
519 7090 CO	3062 ATL F ST	5004 ER 56 FR	6002 MERCU FR
520 FORTR SP	3063 ATL F ID	5005 ER 56 ID	6003 MERCU FR
521 1620 PI	3064 ATLAS LS	5006 ER 56 LP	6004 MERCU FR
522 FORTR SP	3065 ATLAS SF	5007 ER 56 FR	6005 MERCU FR
523 1620T AB	3066 ATLAS ID	5011 2002 DI	6006 MERCU FR
524 1620T GO	3067 ATLAS DP	5012 ALGOL FR	6008 DASK FR
525 1620T DP	3068 ATSYS SY	5014 ALGOL SF	6009 DASK SF
526 1620T ID	3069 ATLAS SY	5015 BULLG LS	6014 FACIT FR
527 1620T LS	3070 ATLAS DP	5016 BULLG DP	6015 FACIT SF
528 FORTR LS	3071 ATLAS DP	5017 BULLG DP	6016 FACIT ID
530 1620 D	3072 ATLAS LC	5018 BULLG LP	6017 FACIT LP
531 1620 D	3073 ATLAS DP	5019 BULLG H	6019 FACIT AB
532 FORTR ID	3074 ATLAS FR	5020 BULLG LC	6020 FACIT AB
533 FORTR TF	3075 ATLAS DP	5021 BULLG LC	6023 FACIT LS
534 7094 FR	3076 ATLAS LS	5022 BULLG SF	6024 FACIT LP
535 FORTR PI	3077 ATLAS ID	5023 BULLG ID	6026 FACIT LF
	3078 ATLAS PL	5024 BULLG PE	
FROM 1ST ED.	3079 ATLAS SP	5025 704 8 FR	NEW, 2ND ED.
3007 SILLI AB		5027 2002 FR	6027 FACIT P
3008 SILLI ID	3512 KDF9 SF	5028 650 DP	6028 FACIT P
3009 SILLI AN	3513 KDF9 PL	5029 650 DP	6029 FACIT AB
3010 SILLI SF	3514 KDF9 A-	5030 650 ST	6030 FACIT E
3012 ZEBRA LI	3515 KDF9 ID	5031 650 ST	6031 FACIT SP
3013 ZEBRA PL	3516 KDF9 DP	5034 Z22 R ER	6032 FACIT SP
3014 ZEBRA P	3517 KDF9 FR	5035 Z22 R SF	6033 FACIT DP
3015 ZEBRA SF	3518 KDF9 ID	5036 650 DP	6034 FACIT DP
3016 ZEBRA MA		5037 650 SF	6035 FACIT SF
3017 ZEBRA SF	FROM 1ST ED.	5038 650 DP	6036 ALGOL H
3018 ZEBRA LS	4001 7090F LS	5039 650 H	6037 ALGOL LS
3019 ZEBRA 1F	4002 704FT 2D	5040 650 DP	6038 ALGOL DP
3021 ZEBRA DP	4007 MERCU LS	5041 650 DP	6039 ALGOL LS
3022 ZEBRA DP	4008 MERCU P	5042 650 DP	6040 ALGOL RE
3023 ZEBRA FR	4011 MERCU DI	3043 650 SC	6041 ALGOL ID
3027 ZEBRA 3	4012 MERCU LC	5044 650 DP	
3028 ZEBRA 3	4013 MERCU 3A	5046 650 DP	FROM 1ST ED.
3029 ZEBRA 3	4014 ATLAS 2S	5047 650 FR	6505 URAL SF
3031 ZEBRA EN	4020 DEUCE ID		6506 URAL DP
3033 ZEBRA SP	4021 DEUCE FR	NEW, 2ND ED.	6507 URAL ST
3035 ZEBRA PI	4022 DEUCE R	5048 ALGOL AB	6512 LGP30 ID
3036 ZEBRA PR	4023 DEUCE DP	5049 ALGOL ID	6513 LGP30 D
3037 ZEBRA DP	4024 DEUCE T/	5050 ALGOL SF	6514 NE803 ID
	4025 DEUCE PL	5051 ALGOL AB	
NEW, 2ND ED.	4028 DEUCE FR	5052 ALGOL LP	NEW, 2ND ED.
3039 PDP 6 SF	4029 DEUCE SF	5053 2002 FR	6515 ZRA1 SF
3040 PDP 6 R3	4030 DEUCE SF	5054 2002 SF	6516 ZRA1 SF
3041 7090 RO	4031 MERCU DP	5055 2002 LP	6517 ZRA 1 LP
3042 FORTR TH	4032 MERCU DP	5056 2002 H	6518 ZRA 1 FR
3043 FORTR LP	4033 MERCU SC	5057 2002 ID	6519 ZRA 1 ST
3044 FORTR DE	4035 MERCU SF	5058 7090 D.	6520 ZRA 1 ST
3045 FORTR ID	4036 MERCU SF	5061 FORTR PA	6521 ZRA 1 ID
3046 FORTR ID	4038 MERCU DP	5062 1620F SP	6522 ZRA 1 SF
3047 FORTR PL	4039 MERCU E	5063 7090 DP	6523 ZRA 1 SF
3048 1620 R	4040 MERCU PA	5064 7090 SF	6524 ZRA 1 SF
3049 ATLAS FR	4041 MERCU R	5065 7090 AB	6525 NE 80 LP

6526 NE 80 ST	7056 PALLA PA	8010 ZEBRA SF	8525 PC1 ID
6527 NE 80 SC		8011 ZEBRA LS	
6528 NE 80 US	FROM 1ST ED.	8012 ZEBRA IT	NEW, 2ND ED.
6529 NE 80 FR	7501 6001 FR	8013 ZEBRA LP	8526 7090 SF
6530 NE 80 SF	7502 6001 FR	8014 ZEBRA LP	8527 7090 LS
6531 NE 80 FR	7503 6001 FR	8015 ZEBRA LP	8528 ALG SF
6532 NE 80 SP	7504 6001 SF	8016 ZEBRA FI	8529 ALG FR
6533 NE 80 LS	7505 6001 SF		8530 7090 LS
	7506 6001 DP	NEW, 2ND ED.	
FROM 1ST ED.	7507 6001 DP	8017 7094 DI	
7001 CAB 5 LS	7509 6001 DP	8018 7094 CO	
7002 CAB 5 LS	7511 6001 SC	8019 7094 EL	
7003 CAB 5 SP	7512 6001 DP	8020 TR4AG PR	
7004 CAB 5 LF	7524 1103 SF	8021 TR4-A 3	
7005 CAB 5 SP	7525 1103 PA	8022 TR4-A LP	
7006 CAB 5 SC	7526 CEP SF	8023 ALGOL SF	
7007 CAB 5 SF	7527 CEP SF	8024 ALGOL LS	
7008 CAB 5 SF		8025 ALGOL PD	
7009 CAB 5 SF	NEW, 2ND ED.	8026 ALGOL SF	
7010 CAB 5 SF	7528 FORTR LS	8027 ALGOL PI	
7011 CAB 5 LP	7529 FORTR LS	8028 ALGOL FR	
7012 CAB 5 ID	7530 FORTR LS	8029 ALGOL SF	
7013 BULLG FR	7531 FORTR SP	8030 ALGOL PI	
7014 BULLG SF	7532 FORTR SF	8031 ALGOL ST	
7015 BULLG D	7533 FORTR SC	8032 ALGOL DP	
7016 BULL LS	7534 FORTR FR	8033 ALGOL ST	
7017 BULL SF	7535 FORTR DP	8034 ALGOL DI	
7018 BULL LP	7536 FORTR LP	8035 ALGOL DI	
	7537 FORTR SC	8036 ALGOL ST	
NEW, 2ND ED.	7538 FORTR ST	8037 ALGOL ST	
7020 FORTR LC	7539 FORTR ST	8038 ALGOL ST	
7021 FORTR 2S	7540 FORTR ST	8039 ALGOL AB	
7022 FORTR N	7541 FORTR ID	8040 ALGOL SP	
7023 FORTR EL	7542 FORTR LS	8041 ALGOL TF	
7024 FORTR PR	7543 FORTR HY	8042 1620 H	
7025 FORTR FT	7544 1620 SF	8043 1620 SF	
7026 7094 S	7545 1620 DF	8044 1620C LF	
7027 7094 FM	7546 1620 DF	8045 1620C FR	
7028 FORTR S	7547 1620 R	8046 1620C SF	
7029 FORTR SP	7548 ALGOL LS	8047 1620C FR	
7030 FORTR FT	7549 ALGOL LS	8048 1620C SC	
7031 FORTR TH	7550 ALGOL SP	8049 1620C SP	
7032 FORTR PI	7551 ALGOL PI	8050 1103 LP	
7033 PALLA LS	7552 ALGOL SF	8051 1103 SF	
7034 PALLA LS	7553 6001 FT	8052 1103 FR	
7035 PALLA LS	7554 6001 FP	8053 1103 FR	
7036 PALLA LS	7555 6001 ID	8054 X1 8 SF	
7037 PALLA LS	7556 6001 LS	8055 X1 8 FR	
7038 PALLA LS	7557 ALGOL DP	8056 X1 8 SC	
7039 PALLA LS	7558 6001 LP	8057 X1 8 DP	
7040 PALLA LS	7559 6001 SC	8058 X1 8 DP	
7041 PALLA LS	7560 6001 TF	8059 ALGOL DP	
7042 PALLA LS	7561 6001 DF	8060 ALGOL ST	
7043 PALLA SP	7562 6001 SF	8061 ALGOL ST	
7044 PALLA SP	7563 6001 AT		
7045 PALLA SP	7564 6001 PA	FROM 1ST ED.	
7046 PALLA SP	7565 6001 AT	8514 M1B SF	
7047 PALLA SP	7566 6001 DP	8515 M1B ID	
7048 PALLA FR	7567 Z23A LS	8516 M1B A	
7049 PALLA ID	7568 Z23A FR	8517 M1B SP	
7050 PALLA SF	7569 Z23A SF	8518 M1B FR	
7051 PALLA SP		8520 PC1 DP	
7052 PALLA SP	FROM 1ST ED.	8521 PC1 SF	
7053 PALLA SP	8007 ZEBRA FR	8522 PC1 DP	
7054 PALLA FR	8008 ZEBRA FR	8523 PC1 TA	
7055 PALLA LS	8009 ZEBRA SF	8524 PC1 D	

APPENDIX -- IBM 360 (FORTRAN 4) PROGRAMS

THE FOLLOWING PROGRAMS ARRIVED TOO LATE TO BE LISTED IN PROPER SEQUENCE WITH
THE OTHER PROGRAMS IN THIS COLLECTION, BUT ARE JUDGED SUFFICIENTLY IMPORTANT
TO MERIT INCLUSION IN THIS APPENDIX --

```
536 360 FT4  D H       PIPPY / AHMED    NRC-21/D FOR POWDER MEASUREMENTS        LWA
    GENERATES THE INDICES AND CALCULATES D(HKL) WITHIN THE SELECTED SPACE
    OMITS PROHIBITED REFLNS. AND SORTS ON D.
537 360 FT4  DIR       HALL / AHMED     NRC-4/SYMBOLIC ADD'N PROCEDURE (C)      LWA
    ESTIMATES THE STRUCTURE FACTOR PHASES OF CENTROSYMMETRIC STRUCTURES
    USING SYMBOLIC ADDITION (SIGMA-2) METHODS.
538 360 FT4  DP        AHMED*,HUBER     NRC-2/DP + GENER'N OF THE LISTS' TAPE   LWA
    GENERATES THE LISTS' TAPE NEEDED FOR STRUCTURE ANALYSIS WITH ALL THE
    NRC PROGRAMS. ALSO REDUCES DIFFRACT'R AND WEISS'G DATA, INTERPOLATES
    ON F-CURVES, ASSIGNS WEIGHTS TO FOBS, APPLIES SHARPENING FUNCTION.
539 360 FT4  FR 3 2    AHMED            NRC-8/FOR DISTORTED+UNDISTORTED NETS    LWA
    COMPUTES THE FR SUMS (A) ALONG LINES PARALLEL TO THE UNIT-CELL AXES,
    OR (B) AT THE GRID PTS. OF A SQ. NET SUPERIMPOSED ON THE SECTIONS OF
    THE UNIT-CELL. THE RESULTS IN (B) ARE PRINTED TO CORRECT MAP SCALE.
540 360 FT4  ID        PIPPY,AHMED*     NRC-12/INTRA-,INTER-MOLR. D + ANGLES    LWA
    SCANS THE GIVEN COORDS. FOR BONDS, DERIVES THE EQUIVALENT ATOMIC POS-
    ITIONS, SCANS FOR INTERMOLECULAR CONTACTS, CALCULATES THE ANGLES, AND
    GIVES SUMMARY OF THE COORDINATION AROUND EACH ATOM. CALCULATES ESD'S.
541 360 FT4  SF LS     AHMED            NRC-10/BLOCK DIAGONAL                   LWA
    REFINES THE COORDS.,ISOTROPIC AND ANISOTROPIC THERMAL PARS., OCCUPA-
    TION FACTOR AND OVERALL SCALE. THE BLOCK SIZES ARE 4X4,5X5,9X9,10X10,
    OR 3X3 AND 6X6 PER ATOM, OR A MIXTURE OF THESE.
542 360 FT4  SPC       PIPPY,AHMED*     NRC-1/THREE CIRCLE GONIOSTAT SETTINGS   LWA
    GENERATES THE INDICES IN ANY DESIRED ORDER OF H,K,L, AND CALCULATES
    THE GONIOSTAT SETTINGS AND 1/LP FOR ALL REFLNS. WITHIN GIVEN LIMITS.
    ASSUMES TWO AXES IN EQUATORIAL PLANE, OR ONE AXIS AT CHI = 90.
```

AHMED,F.R., PURE PHYSICS DIV., NATIONAL RESEARCH COUNCIL, OTTAWA 2, ONT., CANADA
HALL,S.R., PHYSICS DEPT., UNIVERSITY OF WESTERN AUSTRALIA, NEDLANDS, AUSTRALIA
HUBER,C.P., PURE CHEM. DIV., NATIONAL RESEARCH COUNCIL, OTTAWA 2, ONT., CANADA
PIPPY,M.E., PURE PHYSICS DIV., NATIONAL RESEARCH COUNCIL, OTTAWA 2, ONT., CANADA

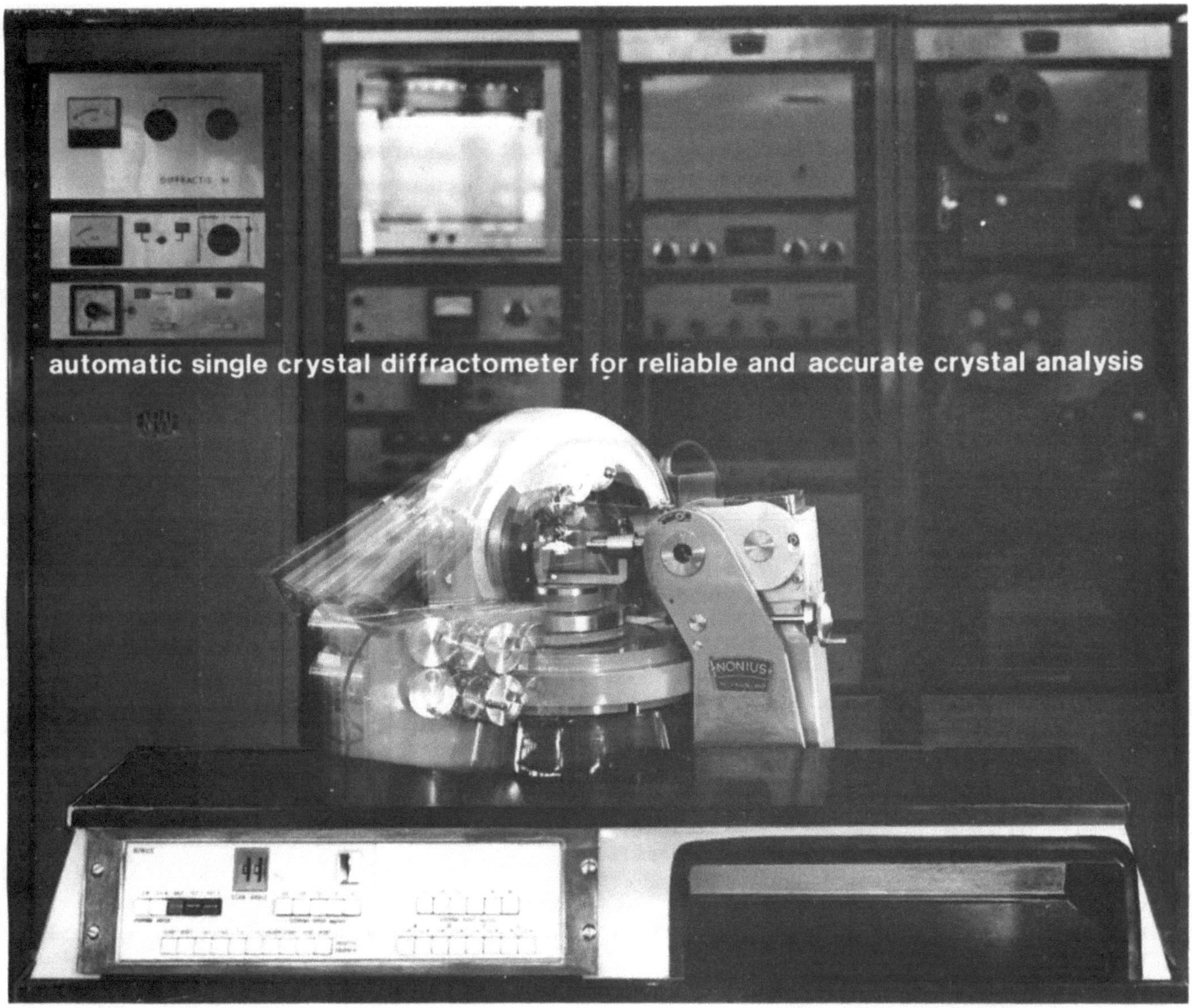

automatic single crystal diffractometer for reliable and accurate crystal analysis

N.V. VERENIGDE INSTRUMENTENFABRIEKEN **ENRAF-NONIUS**

Röntgenweg 1 - p.o.box 83 - telex 31558 - tel. 01730-30950 **DELFT HOLLAND**

INTERNATIONAL UNION OF CRYSTALLOGRAPHY

List of Publications

ACTA CRYSTALLOGRAPHICA

Acta Crystallographica is a scientific journal published in monthly issues and containing original articles in English, French and German on classical and modern crystallography including crystal chemistry and crystal physics.

Subscription price per annum (2 volumes): 400 Danish Kroner, post free ($60 or £21 at the present rates of exchange).

For particulars about prices of back numbers, and preferential subscription rates for *bona fide* crystallographers, please write to the publishers, Messrs Munksgaard, Prags Boulevard 47, Copenhagen S, Denmark.

Various incidental publications of the International Union of Crystallography are mailed at their time of publication, free of charge, to subscribers to *Acta Crystallographica*. In the past Bibliographies on X-ray diffraction at low and high temperatures, a *Crystallographic Book List* and a *World List of Crystallographic Computer Programs* were distributed. Further Bibliographies are in preparation.

Additional copies can be obtained from A. Oosthoek's Uitgevers Mij N.V., Domstraat 11-13, Utrecht, The Netherlands, at the price of 10 Netherlands Guilders (U.S. $3.00, £1 at the present rates of exchange) per copy.

STRUCTURE REPORTS

Vols. 8-21 and 23, covering the years 1940-1957 and 1959, have appeared. Further volumes are in preparation. Vol. 14 is a supplementary volume and cumulative index for 1940-50.

Volume	8	9	10	11	12	13	14	15	16	17	18	19	20	21	23
Years covered	1940-41	1942-44	1945-46	1947-48	1949	1950		1951	1952	1953	1954	1955	1956	1957	1959
Pages	384	448	325	779	478	644	215	588	651	863	846	692	728	706	808
Price in Neth. Glds	80.--	70.--	55.--	100.--	70.--	100.--	35.--	110.--	120.--	125.--	120.--	100.--	100.--	100.--	120.--
*Price in U.S. dollars	22.50	19.50	15.50	28.--	19.50	28.--	10.--	31.--	33.50	35.--	33.50	28.--	28.--	28.--	33.50

For prospectus, order form, and particulars about preferential prices for *bona fide* crystallographers, please write to the publishers, A. Oosthoek's Uitgevers Mij N.V., Domstraat 11-13, Utrecht, The Netherlands.

INTERNATIONAL TABLES FOR X-RAY CRYSTALLOGRAPHY

This successor to *Internationale Tabellen zur Bestimmung von Kristallstrukturen* has been published in three large-size volumes and the publication of additional volumes is being considered.

Vol. 1. Symmetry Groups. Published 1952, pp. xii + 558, price £5.5s.
Vol. 2. Mathematical Tables. Published 1959, pp. xviii + 444, price £5.15s.
Vol. 3. Physical and Chemical Tables. Published 1962, pp. xvi + 362, price £5.15s.

For prospectus, order form, and particulars about preferential prices for *bona fide* crystallographers, please write to the publishers, The Kynoch Press, Witton, Birmingham 6, England.

FIFTY YEARS OF X-RAY DIFFRACTION

This commemorative volume contains the history of the early discoveries, a survey of the development in the various fields of X-ray crystallography, and of Schools and Research in the various countries, and Personal Reminiscences of some thirty crystallographers. x + 733 pp.; price 40 Netherlands Guilders or 11.25 U.S. dollars*.

Publishers: A. Oosthoek's Uitgevers Mij N.V., Domstraat 11-13, Utrecht, The Netherlands.

WORLD DIRECTORY OF CRYSTALLOGRAPHERS

The third edition of this compilation appeared in 1965, and contains biographical information concerning 5037 crystallographers and other scientists from 51 countries. Price 2.50 U.S. dollars*.

Publishers: A. Oosthoek's Uitgevers Mij N.V., Domstraat 11-13, Utrecht, The Netherlands.

* The dollar prices are subject to changes in the official rates of exchange without prior notice.

The publications can also be ordered from Polycrystal Book Service, P.O. Box 11567, Pittsburgh, Pa. 15238, U.S.A., or from any bookseller.

MIX
Papier aus verantwortungsvollen Quellen
Paper from responsible sources
FSC® C105338

If you have any concerns about our products,
you can contact us on
ProductSafety@springernature.com

In case Publisher is established outside the EU,
the EU authorized representative is:
**Springer Nature Customer Service Center GmbH
Europaplatz 3, 69115 Heidelberg, Germany**

Printed by Libri Plureos GmbH
in Hamburg, Germany